A SHEARWATER BOOK

Also by Richard Ellis

The Book of Sharks

The Book of Whales

Dolphins and Porpoises

Men and Whales

Great White Shark (with John McCosker)

Monsters of the Sea

Deep Atlantic

Imagining Atlantis

The Search for the Giant Squid

Encyclopedia of the Sea

Aquagenesis

Sea Dragons: Predators of Prehistoric Oceans

The Empty Ocean

No Turning Back: The Life and Death of Animal Species

Tiger Bone & Rhino Horn

Although this is an herbal medicine and contains no tiger parts, the package suggests that by using it, one can acquire the tiger's potency.

Tiger Bone & Rhino Horn

The Destruction of Wildlife for Traditional Chinese Medicine

Richard Ellis

ISLANDPRESS / Shearwater Books
Washington • Covelo • London

A Shearwater Book
Published by Island Press

Library of Congress Cataloging-in-Publication data.
Ellis, Richard, 1938-
Tiger bone & rhino horn : the destruction of wildlife for traditional Chinese medicine / Richard Ellis.
p. cm.
Includes bibliographical references and index.
ISBN 1-55963-532-0 (hardback : alk. paper)
1. Medicine, Chinese. 2. Poaching. I. Title: Tiger bone and rhino horn. II. Title.
R601.E44 2005
610'.951—dc22 2005002489

British Cataloguing-in-Publication data available.

Printed on recycled, acid-free paper ♾

Design by Joyce C. Weston

Manufactured in the United States of America

10 9 8 7 6 5 4 3 2 1

With a seemingly never-ending demand for the keratinous horn in the Orient, the Greater One-Horned Rhinoceros is facing the biggest threat ever to its existence. . . . This is a testament to the fact that rhinoceros horn, more than any other animal-derived medicine . . . is considered a life-saving traditional Chinese medicine.

Vivek Menon, *Under Siege*

The most serious threat presently to tigers' survival is the use of their bones in Oriental medicine. . . . As wild tiger populations have declined owing to trophy hunting, pest control, and habitat loss, human populations in East Asia have increased dramatically while their per capita expendable income has risen at record rates. At the same time, there has been a resurgence of interest in traditional Asian cures such as tiger bone, the use of which is seen as a status symbol, as a way to retain old customs in the face of rapid change, and as an alternative to the fallibilities of Western medicine.

Judy Mills and Peter Jackson, *Killed for a Cure*

China was the first country to utilize bear bile and gall bladder in traditional medicinal products, and this use was adopted by Korea and Japan centuries ago. Today the use of TCM [traditional Chinese medicine] is widespread not only in Asia but also throughout Asian communities in other areas of the world, including Europe and America. Many of these consumers buy bear bile products, either because they believe it to be a traditional medicine, or because the products are marketed well by local TCM pharmacies. The trade in gall bladders from wild bears has been extensive over the past few decades. Tens of thousands of bears have been killed in the wild to obtain the gall bladders and other parts, including the paws (a delicacy in some Eastern countries), hide, claws, meat, fat and bones. But the gall bladder has been the prize as it has the greatest commercial value.

Tim Phillips and Philip Wilson, *The Bear Bile Business*

Contents

Preface

Nagarahole, India. February, 1997

We are in a small boat on the Kabini River in southern India. It's just me and Stephanie and the boatman. We have come to the Kabini River Lodge to look at wildlife and have already taken several jeep rides into the bush where we've seen elephants, sambar and chital deer, monkeys, monitor lizards, peacocks, and the huge wild ox known as the gaur. The lodge is empty except for us, so the guide, whose name is Sundar Raj, offers to take us out on the river in a small boat after dinner.

We cruise slowly past a large crocodile, immobile on the shore. I take a picture. Sundar cuts the engine so we will make no noise. After about five minutes, he says, "Did you hear that? It was the alarm call of the monkeys. It means a tiger is near." We are silent, listening. A guttural cough. "The tiger," whispers Sundar. We are about 100 feet from shore. We drift closer. The forest trees tower over the narrow strand of the shoreline. An unexpected flicker of movement to the left, and an adult tiger materializes from the forest. It could be a young male, but I decide it's a female. Tigers fear little, and there is no reason for her to look out over the river; either she doesn't see our boat or she sees it and doesn't care. We watch spellbound as she strolls along the strand. I don't want to disturb her with the camera's alien click, and besides, it's getting too dark for photographs. We watch her for what feels like an hour but was probably closer to ten minutes. As darkness falls, she begins to fade, demonstrating how easy it is for a bright orange animal to disappear. Has she reentered the forest or is she still walking? She is gone. We finally exhale. We cheer as if we have found a buried treasure or won the lottery. It is—to me, anyway—the ultimate triumph of tourism. We have traveled halfway around the world and seen temples, dancers, peacocks, giant lizards, wild elephants, even a rhino swimming in the river, but no leopards and, until now, no tigers.

It is difficult to describe the thrill of seeing a wild tiger. Other people in different circumstances or in more light have surely had better sightings. But for me, this event marked the beginning of a yearning for tigers that would not be satisfied until I could write this book—and probably not even then. Although I have in the past written almost exclusively about whales, sharks, and other marine animals, I confess to a secret vice: I love tigers.

When I saw the Kabini River tiger, I was overwhelmed by the great cat's twilight appearance on the strand, and I didn't think for a moment about tiger bones being used in the practice of traditional Chinese medicine. That came later. The tiger was such a living part of its environment that it would have been impossible at that moment to recognize that all tigers—even this one—were endangered by poachers who would kill them and sell their bones to be ground up to make aphrodisiacs and other medicines.

Fish taken from the ocean are typically eaten, more or less in their entirety, but there is one glaring exception: large numbers of sharks of all species are being killed just for their fins to make shark's fin soup. Only the fins are saved; the rest of the shark is discarded. The first dorsal, pectorals, and lower lobes of the tail fin are used in the preparation of this delicacy. The skin is removed, and all muscle tissue as well, leaving only the inner cartilaginous fin rays to soak. The fibers are boiled in water for hours, then the water is changed and the fibers boiled again and again, a process that may take as long as five days. The glutinous mass that results is served in a thick broth of chicken stock seasoned with soy sauce, ginger root, onions, vinegar, mushrooms, and other ingredients. After the soup is drunk, the gelatinous fibers of the shark's fin are eaten.

Shark's fin soup is a prized dish in many parts of Asia and is widely served at Chinese weddings as a symbol of generosity and wealth. As many as forty sharks are killed to supply each wedding, and a bowl of the soup can cost $100. All told, around 100 million sharks are caught each year for their fins, and the hunting has depleted some shark populations by as much as 90 percent. Drinkers of this soup may believe that consuming shark fins transfers to the drinker some of the legendary ferocity of the shark—but it's hard to see why the fins would work that way and not the meat. In many cultures, Western as well as Asian, it is believed that certain foods (or herbs or drugs) can bestow psychological and physical benefits on the consumer far beyond mere nutrition. A great number of plants have indeed been found to have positive medicinal effects—think of the use of quinine, from the bark of the cinchona tree, for the treatment of malaria—and animals (rats, mice, rabbits) are used extensively for medical testing. While Western medicine has moved mainly to synthetics, Chinese medical practice emphasizes the use of "natural" products—animal, vegetable, and mineral—in its 3,000-year-old quest for the cures and prevention of disease.

Practitioners of traditional Chinese medicine often prescribe specific parts of various animals as cures for disease or other malfunctions of the human body or psyche. As we shall see, there are some animals whose various parts are prepared as medicine and whose entire species is endangered by the practice.

This book, like each I've written, grew out of something that I was writing for another. While working on *No Turning Back: The Life and Death of Animal Species*, published in 2004, I was investigating the threat of extinction of rhinos, tigers, and bears in Asia when I realized that many of these species were being killed for their body parts—ingredients for potions used in traditional Chinese medicine. As I later learned, the decline in populations for these species is astounding: where there had been 65,000 black rhinos in East Africa in 1970, there were now 3,600. At the turn of the twentieth century, there may have been 100,000 tigers in India; there were now about 5,000. I wanted to sound the alarm: if the necessary steps are not taken quickly, we will lose forever some of the most charismatic animals on earth. This book is the result.

I am especially indebted to TRAFFIC, an organization that has conducted extensive surveys on the consumption of animals parts throughout the world and summarized its findings in comprehensive reports. I have cited many of these reports in the pages that follow and have acknowledged their authors, but this book could not have been written without TRAFFIC's dedicated field workers, who have made so much critical information available. TRAFFIC—whose acronym stands for Trade Records Analysis of Flora and Fauna in Commerce—is a worldwide organization with headquarters in England that monitors international trade and commerce in wildlife products and is funded in part by the International Union for the Conservation of Nature (IUCN) and by the World Wildlife Fund, whose primary headquarters are in Switzerland.

Because the subject is so far removed from my usual bailiwick, I needed to find people especially knowledgeable about rhinos, tigers, and bears, and I pestered them incessantly with questions, requests for information and papers, and reassurances. Among those who responded to my badgering were John Seidensticker of the National Zoological Park, Washington, D.C.; Peter Jackson and Kristin Nowell of the IUCN Cat Specialist Group; George Amato, Joel Berger, K. Ullas Karanth, and Alan Rabinowitz of the Wildlife Conservation Society (WCS); Jeannie Thomas Parker of the Royal Ontario Museum; Eleanor Milner-Gulland of Imperial College, London; Nigel Leader-Williams of the University of Kent; Rajesh Gopal, director of Project Tiger; Mark Norell and Mike Novacek of the American Museum of Natural History; Jill Robinson of Animals Asia; Rodney Jackson and Darla Hillard of the Snow Leopard Conservancy; Peter Boomgaard of the University of Amsterdam; Clifford Steer of the University of Minnesota Medical School; and Esmond Bradley Martin, Geoffrey Ward, Kees Rookmaaker, Michael Martin, Frank Petito, Kirsten Conrad, and Philip Nyhus.

The results of so much research consultation, together with my almost uncontrollable inclination to quote experts, has probably produced a book that is heavier on quotes than it might have been. I plead guilty to this charge, but with an explanation. I am not a field biologist, and I must depend on the science, descriptions, experiences, and even, in some cases, opinions of others for my text. If I had visited pharmacies in China, tracked moon bears in Thailand, or waited for a glimpse of a snow leopard in the high Himalayas, this would have been a completely different book. Would it have been a better one? Maybe, but I hope I've at least done justice to the investigative work others, and more important, I hope that together we've alerted the world to the plight of the rhinos, tigers, and bears—and the myriad other, less conspicuous species whose parts also make up a large part of the traditional Chinese pharmacopoeia.

Sy Montgomery, a new friend and fellow enthusiast on the subjects of tigers and bears, read the manuscript in an early form, and her suggestions greatly improved what I had written. I sent the bear chapter to Jill Robinson of Animals Asia so that she could read it and correct any of my misinterpretations of her astonishing accomplishments. She carefully read my chapter and made the necessary corrections and updates, and, like the bears she rescued, I am deeply grateful to her. Elizabeth Bennett of the WCS read the manuscript with great care and made more suggestions and corrections than I could have imagined—almost all of which were implemented so I could bring the book up to her high standards of accuracy and professionalism. Like *The Empty Ocean*, this book was edited by Jonathan Cobb, whose deep concern and boundless curiosity about the subject matter—not to mention his superb editing skills—resulted in a much better book than I could have written by myself.

This book, like everything else I do, is for Stephanie.

1

Tyger, Tyger

Wherever wild tigers live, they are revered. Alive they are affiliated with gods or admired for their striking appearance; dead they provide trophies or a pharmacological bounty for consumers. In his book *The Tiger Hunters*, Reginold Burton (born in 1864, so presumably active around 1890) described what happened to the carcass of a tigress that he had shot in India:

> The "lucky bones" were cut out of the chest by Nathu. These are clavicles or rudimentary collar-bones found in all the cat-tribe, about four inches long and hatchet-shaped in the tiger. They are much prized, and, as well as the claws, often mounted in gold and hung round the necks of children to keep off evil. Great care was taken to collect all the fat from the tigress; this was boiled down in a pot over the fire and stored in bottles. The villagers also carried off not only bits of flesh and the liver, but the whole legs and quarters. On being questioned, the Subadar said that the fat was most valuable as a remedy for rheumatism and to make men strong when rubbed into the patient. This is a universal belief throughout the whole of India, where the fat of tigers is everywhere highly prized. He added that the villagers would eat the flesh and especially the liver, the latter being supposed to impart to those who partook of it some of the courage of the tiger.

For centuries, medical traditions throughout Asia have called for the use of exotic-animal parts in healing. However, in the last two decades, the demand for tiger parts has skyrocketed while the population of wild tigers has begun to collapse.

"Beginning about 1986," wrote Geoffrey Ward in *National Geographic* magazine, "something . . . began to happen, something mysterious and deadly. Tigers began to disappear. It was eventually discovered that they were being poisoned and shot and snared so that their bones and other body parts could be smuggled out of India to supply the manufacturers of traditional Chinese medicines."

Until recently, habitat loss was thought to be the largest single threat to the future of wild tigers in India, but while the danger of habitat loss is as great as ever, the trade in tiger bones, destined for use in Asian medicine outside India's borders, is posing an even larger threat. As the Chinese tiger population declined toward extinction, the suppliers or manufacturers of traditional medicines turned to India for tiger bones. The poaching of tigers for the Chinese medicine industry started in India's southern region during the mid-1980s but has now spread to all areas where large number of tigers have been recorded, particularly in the states of Madhya Pradesh, Uttar Pradesh, West Bengal, Bihar, Maharashtra, Andhra Pradesh, and Karnataka. Firearms are sometimes used, but where shooting is impractical, poison or traps are employed. Poison is usually placed in the carcasses of domestic buffaloes and cows; forest pools that serve as water sources may also be poisoned. Steel traps, made by nomadic blacksmiths, are also used and are immensely strong; in a tiger-poaching case near Raipur in 1994, it took six adult men to open a trap. In one area in central India, investigators found that so many steel traps had been set, villagers were afraid to enter that part of the forest.

Major traders operate sophisticated and well-organized supply routes, sometimes to distribute poison, but always to collect tiger bones from even the remotest villages. It is the traders and other middlemen who make most of the substantial profits to be gained from the illegal trade in tiger parts. A tiger can be killed for as little as just over a dollar for the cost of poison, or $9 for a steel trap. Much of the tiger poaching is done by tribal peoples who know the forests well. They are usually paid a meager amount; in a case near Kanha Tiger Reserve, in May 1994, for example, a trader paid four poachers just $15 each for killing a tiger. Sometimes the animals are killed as a result of livestock predation and the body parts are kept in hope of an opportunity to sell them. As the illicit skins move up the commodity chain, from poacher to trader, the potential profits increase exponentially. Although the couriers receive more than the poachers, the real money is made by the traders at the top of the chain, who direct the smuggling syndicates and have links to the buyers. The value of a 2003 Tibet seizure of 31 tigers, 581 leopards, and 778 otters was 6.52 million yuan, or US$787,000. In Tibet, international Environmental Investigation Agency (EIA) examiners were told that a tiger skin was worth US$10,000, a leopard skin was offered for $850, and an otter skin was valued at $250 (Banks and Newman, 2004).

The scale of tiger poaching is enormous considering how relatively few there are left. According to investigations carried out by the Wildlife Protection Society of India (WPSI), a total of 36 tiger skins and 667 kilograms (1,470 pounds) of tiger bones were seized in northern India in 1993–94 alone. In January 2000, officials in Khaga, in the north Indian state of Uttar Pradesh, arrested four people who had 4 tiger skins, 70 leopard skins, and 221 otter skins as well as 18,000 leopard claws, 132 tiger claws, and 175 kilograms (385 pounds) of tiger bones. Up to that time, it was the largest seizure of its kind in India. Because the skins had no bullet holes or snare wounds in them, it was determined that the cats had been poisoned. In the raid on a taxidermy shop south of the city of Lucknow, officials also recovered 1,800 tiger and leopard claws, and 200 skins of the blackbuck, a highly endangered Indian antelope. For the decade between 1994 and 2003, the WPSI documented the poaching and seizure of 684 tigers, 2,335 leopards, and 698 otters in India alone. In *Cat News* (2004), Belinda Wright, Executive Director of WPSI, noted that "Between 12 June and 10 July 2004, 10 tiger skins, 4 sacks of fresh tiger bones, and the claws of 31 tigers and leopards were seized in 11 cases throughout India and Nepal."

The Wildlife Protection Society of India works with government enforcement agencies to apprehend tiger poachers and traders throughout India. It also investigates and verifies reports of unnatural tiger deaths and seizures of tiger parts. The following statistics, supplied by the WPSI, for documented cases of tiger kills in recent years indicate how persistently tiger killings are carried out, though the figures represent only a fraction of the actual poaching and trade in tiger parts in India.

Year	*Number of tigers killed*
1994	95
1995	121
1996	52
1997	88
1998	44
1999	81
2000	53
2001	72
2002	43

The WPSI also tracks reports of tigers described only as "found dead." Many of these deaths are likely due to poaching, but given a lack of clear evidence that this is true, they have not been included in the above figures. Furthermore, since the central and state Indian governments do not systematically compile information on tiger poaching, the WPSI must initially rely on reports from enforcement authorities and other sources, which also underestimate the total. "To reach an estimate of the magnitude of the poaching of tigers in India," WPSI concludes, "it may be interesting to note that the Customs authorities multiply known offenses by ten to estimate the size of an illegal trade." Thus for the years 2000 to 2002, the total number might not be 168, but closer to 1,680.

In a 2004 article about wildlife poaching in the popular magazine *India Today*, Murali Krishnan wrote that not much happens to poachers even when they're caught:

> Besides the inaction in setting up a task force to check poaching, the law too has been lax in imposing penalties or convicting those who have been caught. This has provided poachers the motivation to carry on with their activities unhindered. According to the WPSI, between 1994 and 2003, there were 784 cases of seizure of tiger, leopard, or otter skins. Over 1,400 individuals were accused in connection with these cases, but there were only 14 records of conviction and sentencing.

In 1997, Peter Jackson of the IUCN Cat Specialist Group estimated the total number of tigers left throughout Asia. The most numerous is the "Bengal" (Indian) tiger, with a total population ranging from 3,060 to 4,375. It is found in Bangladesh (300–460), Bhutan (50–240), China (30–35), Myanmar (no information available), Nepal (180–280), and India (2,500–3,750). The Caspian, Javan, and Bali subspecies are extinct. There are also between 437 and 506 Siberian (Amur) tigers, 400 to 500 Sumatran tigers, and 1,180 to 1,790 Indochinese tigers in Cambodia, China, Laos, Malaysia, Myanmar, Thailand, and Vietnam.

Poaching constitutes the most direct threat to India's tigers by targeting ever-dwindling numbers of individual animals, but habitat loss due to growing numbers of humans and their expanding settlements—India's immense population is now over a billion and climbing—exacerbates the problem. When people live in or near places where tigers live, proximity might occasionally end badly for a person, and it almost always ends badly for the tiger. The Panna Preserve, in the central state of Madhya Pradesh, is tropical dry forest, which characterizes some 45 percent of India's tiger habitat. From 1996 to 1997, three researchers, Raghundan Chundawat, Neel Gogate, and A. J. T. Johnsingh (1999), collared and radio-tracked tigers to evaluate their activities and range, and approached feeding tigers on elephant back to see what they were eating. The researchers found that the dry forest habitats support a relatively low population of large ungulates, such as deer and wild pigs, and have a high level of human disturbance. The scarcity of large ungulates means that the Panna tigers have to eat smaller prey such as monkeys to survive, but they are also likely to take cattle, which does not endear them to the local herders. "Tigers in fact take less than 1% of the available cattle each year, but taking cattle on any scale places tigers at the risk of poisoning and creates bad feeling towards them," the authors wrote, urging the creation of areas within the preserve that would be off limits to humans in order to save the remaining tiger population.

Nepal's Royal Chitwan National Park, located just north of the Indian border and incorporating some 360 square miles of the Terai floodplain at the foot of the Himalayas, is the scene of both increased human presence and the threat of poaching. In 1973, fresh out of graduate school (where he had studied radio-tracking mountain lions in Idaho), John Seidensticker first visited Chitwan to collar and track the tigers as a member of the Smithsonian-Nepal Tiger Project. The area had just been designated a park, and the adjacent town of Sauraha (pop. 8,338) was, he observed in his 1996 book, "a sleepy collection of reed houses plastered with cow dung. There were no hotels, inns or restaurants." By 1996, the human population immediately outside the park had grown to 300,000, and the sleepy collection of reed houses had become a tourist center with dozens of hotels, inns, and bars, and viewing platforms from which tourists could watch rhinos and occasionally tigers. There were two lodges, Temple Tiger Camp and Tiger Tops, from which one might venture out on elephant back to see the great cats, as well as rhinos, leopards, various types of deer, monkeys, sloth bears, and blackbuck antelopes.

It's clear that poaching—of tigers and their prey—is reducing the tiger population, but in 1995, John Kenney, James Smith, Anthony Starfield, and Charles McDougal decided to find out by how much. They studied data collected over a twenty-year period in Chitwan National Park and analyzed the survival, fecundity, and dispersal pattern of a population of tigers estimated at between 119 and 210. (As with most tiger populations, those in Chitwan can only be estimated; because the tigers are rarely seen, the estimates are based on comparisons with other populations on the Indian subcontinent, which themselves are unreliable because they are based on the analysis of footprints [pugmarks], which may—or may not—differ from tiger to tiger.) The researchers' computer models showed that "a critical zone exists at which a small, incremental increase in poaching greatly increases the probability of extinction. The implication is that poaching may not at first be seen as a threat but could suddenly become one." In a 1999 follow-up study of Chitwan, another set of researchers, Smith, McDougal, et al., estimated a total of about forty tigers, suggesting that the original estimates of between 119 and 210 were either much too high, or—more likely—that poachers had managed to kill off as many as three-quarters of the Chitwan tigers in four years.

In 1969, when it appeared that the world's tiger populations were already becoming dangerously low, the IUCN held a conference in New Delhi to discuss the problems. Three years later, in conjunction with the World Wildlife Fund, the IUCN initiated "Project Tiger," with the intention of raising support for tiger conservation programs in India. Prime Minister Indira Gandhi set aside nine national parks—Manas,

Palamau, Similipal, Corbett, Ranthambhore, Kanha, Melghat, Bandipur, and Sundarbans—for the protection of India's tiger population, then estimated at about 1,500 animals. India's Project Tiger was set up in 1973, and its first director was Kailash Sankhala. In the early years of Project Tiger, every reserve showed a decrease in hunting and an increase in tigers, leading Dr. Sankhala to write in 1977, "I am greatly encouraged by the response of the habitat, the tigers, and their prey in the Tiger Reserves. It may be too early to predict the outcome of this effort, but it is surely not too much to hope that ultimately the tiger will be restored to a less precarious position than he is at present."

Alas, it was not to be. The tiger is now in a *more* precarious position than he was in 1977. Since the inception of Project Tiger three decades ago, the population of India has increased by 300 million people and livestock numbers have risen by 100 million. Indira Gandhi was assassinated in 1984, and without her support, Project Tiger all but evaporated, largely because it was not given the resources required to be effective. Although the project still exists—and now oversees another eighteen reserves—almost all the reserves were invaded by settlers who needed food and fodder, and who regarded the tigers as a nuisance or, in some

cases, a threat. Tigers were killed because they interfered with farming, and some because they interfered with the very lives of the farmers. And starting in the late 1980s, they were increasingly killed for use in traditional Chinese medicine.

In a report published in 1998, the director of Project Tiger, P. K. Sen, commented on these tiger killings (as quoted in Valmik Thapar's *The Secret Life of Tigers*):

> It is my considered opinion, after more than one and a half years as Director [of] Project Tiger, that the tiger and its ecosystem is facing its worst ever crisis. I feel that the figures of one tiger death every day may even be an underestimate and there are many reasons to say so. If one out of every ten tigers poached, poisoned or crushed under the wheels of a vehicle, three are tigresses who have cubs, all the cubs will die unnoticed because they are totally dependent on the mother. The death of three resident male tigers will result in new males occupying the vacant ranges, and in the first instance they will kill all cubs in order to father their own litters. Thus for every ten tigers killed, sometimes as many as fifteen additional tigers die. . . . On the occasion of 25 years of Project Tiger, unless revolutionary steps are taken immediately, there is little hope for the future and we could be reaching the point of no return.

When Richard Perry published *The World of the Tiger* in 1964, tigers had been hunted for centuries throughout their range, and only the Balinese tiger was known to have been driven to extinction. The Caspian tiger would be seen for the last time four years after Perry's book appeared, and the Javan tiger was still hanging on. Though the world's tiger population was only about 15,000, according to Perry, few people had begun to worry that the entire species would become extinct. Perry himself, however, was all too aware of the unresolved conflict between human and animal populations, and he wrote: "One suspects that in the end it will probably be solved in a manner disastrous for both men and animals; but we must continue trying to solve it humanely."

Humans are not to be defined by their humane behavior, especially to animals. In the decades following publication of Perry's book, the

Javan tiger disappeared and tiger populations seemed to be falling everywhere in India, so "Project Tiger" was initiated to protect them. Just when it looked as if they might recuperate, however, poachers began killing them with such celerity that the remaining Indian tigers have careened toward extirpation. White hunters and maharajas behaved as if their goal was to eliminate the tiger by their wanton hunting, and in their wake, the poaching brigade, attracted in large part by the lucrative trade in animal parts for traditional Chinese medicine, is nearing dubious success in wiping wild tigers off the face of the earth.

Accidentally or intentionally, the conflict between human and animal populations that Perry referred to is leading to the imminent extinction of many species. Habitat destruction, hunting, fishing, pollution, global warming, and numerous other factors have been identified as significant threats to the world's wildlife. The demand of traditional Chinese medicine for animal parts is emblematic of this conflict. The reasons for killing bears, rhinos, elephants, seals, sea horses, and numerous other species are sometimes different than the reasons for killing tigers, but extinction is extinction, no matter what the rationale or explanation.

2

Suffer the Animals

At another stall, we found the head of a Hoolock's gibbon alongside a shoulder bag made from the animal's torso and arms. Its brain was good for headaches, the owner said. Then he showed us more animal parts and pointed me in the direction of other stalls selling wildlife. Within the first hour I had a species list that included parts ranging from Himalayan black bear, black serow, wild dog, and leopard cats to some of the rarest, most unusual species in the world—takin, musk deer, red goral, and red panda.

Alan Rabinowitz in Putao, Myanmar (Beyond the Last Village, 2001)

Over time, humans developed a very special relationship with the so-called higher vertebrates (the warm-blooded birds and mammals): some we domesticated, others we killed off in vast numbers. There are those who argue that we killed off *all* the "charismatic megafauna" of the North American Pleistocene: the mammoths, mastodons, wooly rhinos, saber-toothed cats, mega-hyenas, and almost every other large mammal species that lived in North America thirteen thousand years ago. (Not everybody believes this; some think that climate change or disease did them in, or at least were strong contributing factors, but whatever the cause, they are extinct.)

In more recent times, *Homo sapiens* the not-so-wise has succeeded in eliminating the dodo, the passenger pigeon, the Carolina parakeet, the great auk, the ivory-billed woodpecker, Steller's sea cow, and dozens of other less-familiar animals. In its wisdom, the Australian government condoned and even encouraged the eradication of the Tasmanian wolf (*thylacine*), because officials believed (wrongly) that these marsupial carnivores were a threat to sheep, but the extinction of the species was not intentional policy. Nobody in authority decided that it would be a good idea to eradicate the passenger pigeon, at one time the most numerous bird in North America. The number of birds and mammals on the "endangered" list is enormous, and the list grows longer every day. Huge numbers of animals have been killed for commerce; if we didn't kill off all the fur seals and sea otters for their luxurious coats, or the whales for their baleen and oil, it was not for want of trying.

For some reason, very large terrestrial mammals with strange protrusions from their faces have always been targeted by collectors and in some instances avidly sought for what were believed to be their medicinal properties. The rhinoceros has "horns" growing atop its nose, while some other large creatures are blessed with huge ivory teeth. Those with the most prominent teeth are elephants, walruses, and the narwhal, a little Arctic whale with a single ivory tooth that projects straight out in front of it, which would play an important part in the story of the fabled unicorn. In *Love, War, and Circuses*, Eric Scigliano's 2002 book about humankind's relationships with elephants, we find this vivid description of ivory's nonmedical qualities:

> Unlike useful meat and hide, ivory holds largely aesthetic and symbolic value. Although it has been used throughout the ages to make utilitarian objects ranging from arrowheads and fishhooks to spoons and horns, it is not the only or even the best material for any. It is harder and much harder to work than clay or wood, but not so hard or readily sharpened as stone and metal. It cannot be molded, bent, cast, or flaked into shape; it must be painstakingly carved. But for carving it is matchless, taking detail like no other substance, glowing with an inner luster that deepens with age, warm as wood and fine as metal.

"The tusks fetch a vast price," wrote Pliny, "and supply a very elegant material for images of the gods." The Greek sculptor Phidias made a 40-foot-high statue of the goddess Athena for the Parthenon (the *Athena Parthenos*), which was dedicated to her in 438 BC. The statue, like Phidias' earlier *Zeus*, which is now considered to be one of the Seven Wonders of the Ancient World, was a colossal figure of chryselephantine workmanship—draperies of beaten gold, flesh parts encrusted with ivory. To the Romans, ivory symbolized wealth and power. The mad emperor Caligula (AD 12–41) even kept his favorite horse inside the palace in a stable box of carved ivory, dressed in purple blankets and collars of precious stones. (The horse, who was made a senator, was often invited to dine with the emperor.)

With the fall of the Roman Empire, the flow of ivory into Europe was reduced to a trickle, largely because the elephants had been eliminated from North Africa and Asia Minor. The scarcity of elephant ivory made it that much more desirable, and in the Middle Ages it was a popular material for combs, mirror backs, chessmen, small boxes, jewelry, religious icons, small caskets (known as reliquaries), carvings of the Virgin, and the spectacular twelfth-century cross of Bury Saint Edmunds. Ivory was used in China for decorative objects, fans, chopsticks, and model houses and boats; and in Japan for the small toggles known as *netsuke* and *okimono*, which are larger carved sculptures, the latter most often of a human form. Elsewhere, ivory was used extensively in the manufacture of piano keys and billiard balls. Ivory was also collected from already-dead mammoths frozen into the tundras in Alaska, Canada, and Siberia, but nothing could match the great tusks of the elephants of Africa, which were slaughtered in unimaginable numbers for the ivory trade and carried to the coast by captives who were then themselves sold as slaves.

There is no way to obtain elephant ivory without killing its original owner. As elephants became scarcer, so too did their ivory, and carvers and manufacturers had to look elsewhere for approximations of the creamy white substance so familiar that it has given its name to a color. Although they are not as large or as fine as the ivory from elephants, the canine teeth of hippopotamuses, the tusks of walruses, the spiral tooth of the narwhal, the peglike teeth of the sperm whale (from which authentic scrimshaw is derived), and even the bones of large mammals such as cows and horses have been carved into "ivory" objects. The casque of the large tropical bird known as the helmeted hornbill (*Buceros vigil*) can be carved into small objects, and the inner lining of the seed of the tagua palm (*Phytelephas macrocarpa*) can be sliced like real ivory and made into ivory-like buttons.

Elephants have been used as beasts of burden and as sources of ivory for centuries, but they also play a part in traditional Chinese medicine. As with many other creatures—rhinos, particularly—that are not found in China, suppliers for traditional medications had to range far afield for their sources. In Li Shih-chen's 1597 materia medica* *Pen Ts'ao Kang Mu* (Compendium of Materia Medica), one of the fundamental texts for Chinese medicine, we find the Indian elephant prominently listed as a source of various medicinal potions. For example, the tusks, when ground or pared, are "sweet, cooling, and nonpoisonous," and can be used to remove foreign bodies from the throat and to cure epilepsy, osteomyelitis, and smallpox. Eating elephant flesh will make a person heavy. Elephant bile will clarify the vision and can be applied to the gums at night for halitosis. Ashed elephant skin can be applied with oil to heal injuries from metal weapons. Elephant bone is a general antidote to poisons; the bone ash is good for weakness of the stomach, heartburn, vomiting, cholera, diarrhea, colic, and poor appetite. The small horizontal bone in the chest, ashed and taken with wine, will cause a man to float.

The use of elephant products for medicinal purposes did not end in the sixteenth century, when some people floated and others clarified their vision with elephant bile. It goes on today, as described in a 2002 TRAFFIC report by Caitlin O'Connell-Rodwell and Rob Parry-Jones. Ivory powder collected from carving factories is still sold in pharmacies and used to warm the liver. We do not know how much elephant skin is traded without detection, but the skin of twenty elephants was seized by officials in Yunnan Province in 2001, and later that year, another 15 tons—for which some 260 elephants died—was seized by the Guangzhou Forestry Police. Modern uses are not confined to bones and skin. Cited in Alan Rabinowitz's *Chasing the Dragon's Tail* is a *Bangkok Post* article of November 11, 1987, describing "the discovery of nearly a dozen elephant carcasses in the forests of Kanchanaburi [Thailand], a province adjacent

* *Materia medica*, literally translated, is "medical material." The term is used to describe drugs and other substances that physicians prescribe to cure illness. A *pharmacopoeia* is a book containing a list of drugs and other medicinal substances with directions for their preparation and identification, but it can also be a stock of these substances.

to the Burmese border. Their penises had all been cut off. Elephant penises, which can weigh nearly forty-five pounds, are worth nine to ten dollars a pound on the black market, and are purchased by the Chinese as an aphrodisiac."

Tusks also characterize another animal that figures in traditional Chinese medicine, but the tusks have no medical value. If you were to find a musk deer skull in an Asian market, unless you were a zoologist you might think it belonged to a miniature saber-toothed tiger. Of course, it would be altogether the wrong shape and would not have the cat's carnassial (shearing) teeth, but 4-inch tusks on a 2 $\frac{1}{2}$-foot-tall deer are more than a little odd. Both males and females have tusks, but neither sex has antlers. With tusks and no antlers, deer of the genus *Moschus* (there are five poorly defined species) are so unusual that it is not surprising to learn, as Valerius Geist informs us in *Deer of the World*, that "musk deer climb trees and cliffs." With its rounded, fuzzy ears, a musk deer's head looks remarkably like that of a kangaroo. One species or another is found in the high-elevation forests and brushlands of Siberia, India, Pakistan, Vietnam, China, Mongolia, Korea, and Myanmar. Habitat destruction has been responsible for a crash in all musk

Musk Deer (*Moschus moschiferus*).

deer populations, but far more pernicious has been the harvesting of the deer for the gland that gives them their name. Under the adult male's tail is a walnut-sized gland that secretes a yellowish, waxy substance that is believed to influence the estrous cycle in females and therefore plays an important part in the deer's reproductive cycle. The musk, which has a powerful and penetrating odor, has been employed for centuries in the manufacture of perfumes—more for men than women. Even today, many colognes and other scents for men emphasize "musk," but it is more than likely to be artificial, because the real thing is so expensive. Musk is reputed to be the most expensive animal product in the world, "fetching up to $65,000 per kilogram [2.2 pounds] on the international market" (Macdonald 2001).

In the traditional Chinese medicine (TCM) pharmacopoeia, musk has always played an extremely important role. In Li Shih-chen's 1597 materia medica, we read that the musk can be used "for acute pain and swelling of the abdomen, for constipation, difficult labor . . . for snake and rat bites. For caries . . . For restoring a lost sense of smell, for opening up the functions of the body orifices." And that was in the sixteenth century. In Bensky and Gamble's 1993 *Chinese Herbal Medicine Materia Medica*, musk is a miracle drug that "opens the orifices, relieves the spirit, and unblocks closed disorders; because of its intensely aromatic, penetrating nature this substance is used in treating a wide variety of problems that impair consciousness . . . Invigorates the blood, dissipates clumps, reduces swelling and alleviates pain. " In his 1999 TRAFFIC report on the musk deer, Volker Homes observed that, aside from its centuries-old use in the West for perfume, "the most important market for musk products is now Asia, for traditional East Asian medicine. Musk is included in about 300 pharmaceutical preparations in traditional Chinese and Korean medicine as a sedative and a stimulant, to treat a variety of ailments of the heart, nerves, breathing and sexuality and is therefore one of the most commonly used animal products in this type of medicine."

According to a 1998 TRAFFIC report by Judy Mills, the annual demand for musk in China alone is estimated to be 500–1,000 kilos (1,100–2,200 pounds), which would require the glands from at least one hundred thousand deer. While it has been estimated that some seven hundred thousand musk deer remain in the wild, no one knows how long these species can withstand the current levels of hunting to meet commercial demands. It is possible to remove the musk gland surgically, which allows the deer to survive, and there are even musk deer farms in China, but for the most part, populations of the deer are hunted and killed in substantial numbers, and they are now endangered—or in some cases, extinct—wherever they live (or lived). Homes, for example, tells us that Russia's musk deer populations have been halved in the last decade as a result of poaching and illegal trading.

Animals that are killed illegally are said to be "poached," a term defined by the *Oxford English Dictionary* as "to catch or carry off (game or fish) illegally; to capture by illicit or unsportsmanlike methods such as a poacher uses." In many regions, wild animals are killed for food, and it is more than a little awkward to raise the issue of poaching where hungry people have to hunt to eat. In the modern trade for animal parts, the line between poaching and food-hunting is often blurred, but both are deadly. (Much noncommercial subsistence hunting is sustainable, of course, and does not threaten endangered species.) Joel Berger, in *Horn of Darkness*, the story of his rhino research in southwest Africa, explains:

> The fact that people are famished does not justify breaking laws, but it is far easier to understand that type of motivation than large-scale commercial poaching operations, often run through organized gangs. A barter economy, including the trade and sale of wildlife products, has gone on for centuries. Today's story is different. Modern weapons coupled with the lure of money from foreign markets drive unscrupulous traders to exploit impoverished people. The result has been the decimation of animal populations. Ostrich feathers were once fancied as ornaments in Europe and the Americas. Gall bladders and tiger bones are used medicinally in Asia.

The scope and breadth of interest in animal parts for Chinese medicine is staggering. In Bernard Read's 1931 translation of the sixteenth-century Chinese materia medica *Pen Ts'ao Kang Mu*, Li Shih-chen incorporated four categories of "Animal Drugs": domestic animals; wild animals; rodentia, monkeys, and supernatural beings; and man as a medicine. Among the domestic animals are pigs, dogs, horses, asses, cows, and geese; listed with the monkeys are gibbons and orangutans, as well as sea horses, sprites, dryads, and naiads; and those parts of humans that can be used medicinally are hair, bones, placenta, bile, penis, and umbilical cord. The wild animal list is the longest and includes lions, tigers, leopards, tapirs, elephants, rhinoceroses, yaks, wild boars, porcupines, bears, deer, elk, civets, badgers, rabbits, otters, and wolves. Lions, elephants, and tapirs are not found in China now and probably were not in the sixteenth century, so they had to be harvested elsewhere, and sea horses, those tiny fishes that swim upright and look like miniature horses, are so strange looking that they seemed a natural for inclusion in the traditional Chinese zoomorphic pharmacopoeia.

There are some thirty-five species of sea horses, ranging in size from the tiny pygmy (*Hippocampus zosterae*), the mature adults of which are less than a half-inch long, to the 8-inch-long *H. ingens*. While a single pygmy sea horse wouldn't make much of a meal for a kitten, thousands of them would do, even for people. In her 1994 article in *National Geographic*, Amanda Vincent, an authority on sea horses, wrote that she had found "wok-fried-sea horses" on a menu in Hobart, Tasmania. Dried sea horses are used to make key chains, jewelry, paperweights, Christmas tree ornaments, and other souvenirs. Sea horses found in Indonesian and Philippine waters are used also in these island nations for folk medications. Shrimp trawlers using small-mesh nets often collect sea horses as a bycatch, and these little creatures ultimately find their way into apothecaries or souvenir shops.

By far the largest number of sea horses harvested are shipped to China (and to a lesser extent, to Korea and Japan), where practitioners of traditional Chinese medicine have been using sea horses for the past five centuries to cure impotency and asthma, lower cholesterol, prevent arteriosclerosis, and even enhance virility. In *Fundamentals of Chinese Medicine* (Zhōng Xī Xué Jīi Chŭ), the 1996 Wiseman and Ellis translation of a textbook used in several Chinese medical schools, the entry for "sea horse" (*hăi mă*) reads as follows:

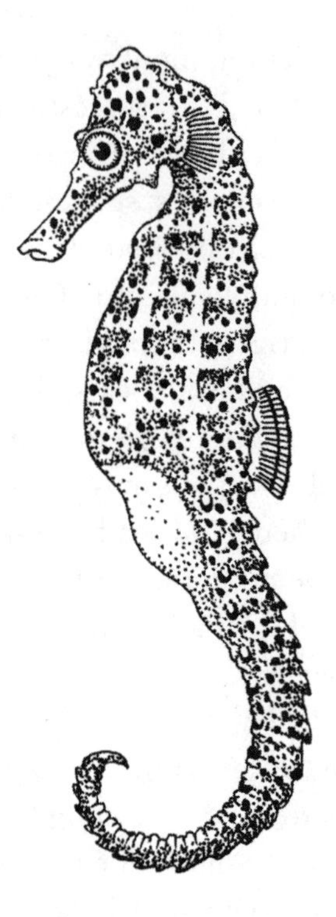

Sea horse (*Hippocampus* sp.).

> Warm, sweet, nontoxic. Enters the liver and kidney channels. Supplements the kidney and strengthens yang, regulates qi and quickens the blood. Treats impotence, enuresis, vacuity panting, difficult delivery; conglomerations and accumulations; toxin swelling of cold sores. . . . *Caution*: Contraindicated in pregnancy and effulgent yin vacuity fire.

The 1986 *Chinese Herbal Medicine: Materia Medica*, as translated by Dan Bensky and Andrew Gamble (1993), adds: "Tonifies the kidney and fortifies the yang; for impotence, urinary incontinence, deficiency wheezing, and debility in the elderly. Invigorates the blood: for bleeding and pain from blood stasis, and swelling due to sores and boils. Also used for abdominal masses. . . . Often steeped in wine when treating impotence. Good quality is whole, big, and firm."

Amanda Vincent and her colleagues at Project Seahorse have been instrumental in bringing to light the plight of sea horses and the trade that has developed around them. "Vincent is an energetic advocate for a beguiling but beleaguered group of fishes that derive their name from their startling resemblance to horses," wrote Floyd Whaley in the August 2004 issue of *Wildlife Conservation*. "As director of Project Seahorse, she has almost singlehandedly put seahorses and their relatives on the radar screens of the world's scientific and conservation communities." The size of the world's sea horse population remains a mystery, but it is difficult to imagine a population of animals—even if it is composed of numerous species—that can withstand the removal of 20 million per year, but that appears to be the number taken annually. In her Project Seahorse report, Vincent is quoted as saying that "in the year 2000, more than 20 million seahorses were traded globally, a dramatic increase from years past. With mainland China's growing economy and population, demand will probably climb."

Project Seahorse estimates that "total global consumption of seahorses was at least 21 million seahorses in 2001 (more than 60 metric tons) . . . [but] this now appears to be an underestimate; new and very incomplete data from Hong Kong show imports of nearly 13.5 tons from that region alone." In the Web pages entitled "Seahorse Biology and Conservation," Vincent and Hall summarize what is known (and not known) about the sea horse population:

> Extracting seahorses at current rates appears to be having a serious effect on their populations. The impact of removing millions of seahorses can only be assessed indirectly because global seahorse numbers are unknown, taxonomic identities are unclear, geographic ranges are undefined, and fisheries undocumented. Nonetheless, most participants in established seahorse fisheries reported that catches were dwindling markedly.

Other sea creatures have also played a part in the traditional medical practices of China. In his 1874 book, Captain Charles Melville Scammon, who was a sealer before he turned to whaling (and later opposed the killing of seals and whales), wrote about a herd of sea lions that were killed on Santa Barbara Island in 1852:

> The herd at this time numbered seventy-five, which were soon dispatched, by shooting the largest ones, and clubbing and lancing the others, save one Sea Lion, which was spared to ascertain whether it would make any resistance by being driven over the hills beyond. The poor creature only moved along through the prickly pears that covered the ground, when compelled by his cruel pursuers; and at last, with an imploring look and writhing in pain, it held out its fin-like arms, which were pierced with thorns, in such a manner as to touch the sympathy of the barbarous sealers, who instantly put the sufferer out of his misery by the stroke of a heavy club. As soon as the animal is killed, the longest spires of its whiskers are pulled out, then it is skinned, and its coating of fat cut in sections from the body and transported to the vessel, where, after being "minced," the oil is extracted by boiling. The testes are taken out, and, with the selected spires of the whiskers, find a market in China—the former being used medicinally, and the latter for personal ornaments.

In California in the early 1900s, Steller's sea lions (*Eumetopias jubatus*) were killed in large numbers because fishermen complained that the sea lions were eating the fish that belonged to the fishermen, and later, in the Aleutians, pups were killed for their skins to make clothing. Most male pinnipeds (seals, sea lions, and walruses) have a baculum or penis bone, and it is not surprising to learn that these bones were ground up and the powder sold as an aphrodisiac in certain Asian countries. In *The War Against the Seals*, an aptly named study of North American sealing, Briton Cooper Busch discusses "trimmings," which were composed of the genitals of bull seals, the gall bladder, and the whiskers. The penis and testicles were powdered and used to impart virility to men, the gall bladder was used for medicine, and the whiskers made wonderful toothpicks or opium-pipe cleaners. The whole "set"—genitals, gall bladder, and whiskers—brought $2 to $5 in San Francisco's Chinatown. Because the pelts were not particularly useful, the animals were often killed just for the "trimmings" and the carcasses left to rot on the beach.

Steller's sea lions are "ursine" (bearlike) seals, as are northern fur seals (*Callorhinus ursinus*). Although they look quite different, their ranges overlap, particularly in the western North Pacific, and those who killed them for their "parts" probably didn't care which was which. Seal and sea lion parts are still being used in traditional Chinese medicine. *Rare Chinese Materia Medica*, published by the Shanghai University of Traditional Chinese Medicine in 1989, more than a century after Scammon's discourse on seals' testes, contains the following:

> PREPARATION
>
> Cleaned ursine seal's penis and testes are dried and ground into powder, or they are moistened with liquor first, then roasted and ground into powder.
>
> NATURE, TASTE AND CHANNEL TROPISM
>
> The medicine is salty in taste and hot in nature. Its therapeutic action is released to the channels of the liver and kidney.
>
> ACTIONS AND INDICATIONS
>
> The medicine can warm the kidney to invigorate *yang* and replenish the vital essence to reinforce the marrow. It is efficacious in the treatment of consumptive disease, impotence, cold sperm, lassitude in the loins and legs due to the deficiency of the kidney-*yang*, etc.

Wiseman and Ellis (1996) on "Seal's Genitals" (*hǎi gǒu shèn*): "Hot; salty; nontoxic. Enters the liver and kidney channels. Warms the kidney and strengthens yang; boosts the essence and supplements marrow. Treats vacuity detriment taxation; impotence; limp wilting (*wěi*) lumbus and knees." In *Chinese Herbal Medicine* (1993), Bensky and Gamble suggest that for best results, seal penis should be combined with the fungus *Cordyceps sinensis*. In Chinese, this substance, *dōng chóng xià cǎo* ("winter-worm, summer-grass"), is actually the fruiting body of a parasitic fungus that invades the bodies of caterpillars and eats the soft tissue. One TCM Web site selling *Cordyceps* says, "In Traditional Chinese Medicine, *Cordyceps* has long been used as an aphrodisiac. Recent research in China and Japan has shown a 64% success rate among men suffering from sexual dysfunction, vs. 24% in the placebo group." In his 1987 article on wild animal products used as aphrodisiacs, Esmond Bradley Martin comments:

Steller's Sea Lion (*Eumetopias jubatus*).

> Asians consume incredible amounts and varieties of animal products and herbs for medicinal purposes, and they consider some especially useful for enhancing their love lives . . . seal pills—made from top-quality testicle and penis of seals, ground into a powder and pilled in form convenient for intake, stimulate male virility. (1987a)*

Evidently, the mutilation of seals in North America for pharmacological purposes did not end in the nineteenth century either; it is still going on. In an article in the *Portsmouth Herald* for October 30, 2003, Karen Dandurant reported on five harbor seals that were found skinned and mutilated on New Hampshire beaches. Dandurant quotes National Marine Fisheries Service agent Chris Shoppmeyer:

> They were professionally skinned, with no damage to the internal organs. Whoever did this was skilled with a knife. As to the other

* Martin wrote this in 1987; Western pharmacology has now also provided a treatment for "erectile dysfunction" in the form of Viagra and other prescriptions.

> mutilation, there is a big black market for the male genitalia of seals. In some Asian countries it is thought to be a powerful aphrodisiac, bringing in several hundred dollars per ounce. . . . The market for wildlife and their body parts is second only to the drug trade. . . . If I catch these guys [the cases] are going criminal. We're seeing a clear intent of commercialization.

Seal genitals, tiger bones, rhino horn, and the musk gland of a little deer all play a role in traditional Chinese medicine. We tend to think of scales as comprising the integument of fishes, snakes, and lizards, so the pangolin, a *mammal* with scales, is weird enough to earn a place in the TCM pharmacopoeia. There are seven species of pangolin, also known as "scaly anteaters": four in Africa and three in southern Asia. Among the African pangolins is the giant, *Manis gigantea*, which can measure six feet from the tip of its nose to the end of its thick, scaly tail, but most species are smaller, less than three feet overall. They exist almost exclusively on ants and termites, which they harvest with their incredibly long, sticky tongues, after digging up the anthills with their powerful foreclaws. (Anteaters too have long, sticky tongues, powerful foreclaws, and no teeth, but pangolins and anteaters are not closely related; their similarities are attributable to convergence, the evolutionary process in which unrelated species develop similar modifications that allow them to function more or less in the same fashion.) Some pangolin species live in ground burrows, whereas some are arboreal and shelter in hollow trees; the tree-climbing pangolin can hang from branches with its prehensile tail. When disturbed, pangolins can roll up in a tight ball, their scales providing an almost impervious defense against predators such as leopards, and they are also capable of exuding a foul-smelling liquid from their skin.

Not many animals would try to unroll a smelly, heavy-scaled pangolin, but *Homo sapiens*, the most efficient predator the world has ever known, has figured out that pangolins can be unrolled when they're dead, and the very scales that would have protected the pangolins in life are the main reason they are being killed. In a 1937 report, G. A. C.

Chinese pangolin (*Manis pentadactyla*)

Herklots wrote, "A stranger to Hong Kong might wonder why this harmless, inoffensive nocturnal creature, which lives largely on 'white ants,' the worst pest of the tropics, should need protection. Unfortunately for the animals, however, the Chinese believe that its scales have remarkable medicinal properties, and the animal itself is also eaten." Another of Bernard Read's twentieth-century publications of Li Shih-chen's 1597 *Chinese Materia Medica* was devoted to "Dragons and Snakes," one of which was the pangolin. (Others were crocodiles, lizards, and whales.) Pangolin scales were prescribed for "excessive nervousness and hysterical crying in children. For women possessed by devils and ogres. . . . Given with aconite and oyster shells for paralysis of the hands and feet; scales from the right side of the animal are given for affections [sic] on the left side of the body and vice versa." Ash from a big piece is "mixed with oyster shell, seven scorpion's tails, a little musk, linseed oil, and wax to form a small rod which is wrapped in cotton and rammed into the ear for ringing sounds and deafness due to sexual weakness. . . . For two months the powdered ash blown into the nose while the patient holds water in the mouth is used to cure eyelashes which curve inwards."

In modern TCM, the scales are still alleged to be powerful medicine, and the 1989 *Chinese Materia Medica* of Zhang Enquin tells us that the scales, "washed clean, dried in sunlight, stir-baked with sand, soaked in vinegar and dried for use," can be ground into a powder that will promote blood circulation, stimulate milk secretion, subdue swelling, and ease sores, carbuncles, and skin infections. Because pangolins figure prominently in Southeast Asian traditional medicine, it is convenient that three species are found in the very area where their scales can be utilized. "From 1958 to 1965," according to Ronald Nowak (1991), "over 60 tons of scales were exported legally from Sarawak, mainly to Singapore, and are estimated to have represented over 50,000 animals that were taken almost entirely in Kalimantan (Indonesian Borneo). . . . From 1980 to 1985, an estimated 3,000 to 5,000 pangolins were imported annually to Taiwan and South Korea." More recently, in 2002, some 600 pangolins were seized at Hanoi's airport after being illegally smuggled in from Malaysia en route to China, and the following year, more than 6,600 pounds of pangolins and more than 1,100 pounds of tortoises were discovered in 240 cargo cases from a Singapore Airlines flight. In the 1990s, the high volume of largely unregulated trade, including illegal shipments, brought about a measure of protection: all Convention on International Trade in Endangered Species of Wild Flora and Fauna (CITES) parties decided to adopt a "zero quota" for Asian pangolins, which effectively bans all international trade. Since 2000, all trade in Asian pangolins has been illegal.

Many rare animals are killed for the pharmacological requirements of traditional Chinese medicine. The valuable gall bladder and bile of endangered bears are used to treat a variety of inflammations, infections, and pain. There are "bear farms" that supply bile to Chinese practitioners, where steel catheters are surgically implanted into caged bears' gall bladders, enabling handlers to regularly "milk" the bears for their bile. North and South American bears are also being killed for their gall bladders, which are then smuggled into China. Other endangered species, such as the various rhinos, are killed for their horns, which, when ground into a powder, are said to cure diverse ailments. Tigers, endangered throughout their range, are killed for their bones, which are made into a "tiger bone wine" said to give the drinker the strength of the tiger. Tiger penis-bone soup is thought to be an aphrodisiac; eyeballs rolled into pills are thought to cure convulsions; whiskers protect one against bullets. Musk deer are killed for a special gland. Russian hunters have decimated huge herds of saiga antelope for their horns. Tons of dried sea horses enter the TCM trade each year. Throughout the world's oceans, sharks of all species are being slaughtered and their fins cut off to make ridiculously expensive ($100 a bowl) shark's fin soup, which is believed to confer the savagery of the shark on the drinker. Ever since people developed the notion that the horn of the unicorn could detect poison or cure various ailments, people have killed animals to obtain "medicine" from various parts. Modern medicine, although far from available worldwide, still provides cures for many diseases and infirmities. It is a terrible anachronism that so many people today rely on largely ineffectual animal-related remedies, but the real tragedy is that large numbers of animals have to die to provide these nostrums.*

The great majority of medicinals prescribed in TCM are of a vegetable or herbal origin; only a few originate in animal parts. Of these, many are from domestic animals such as pigs, cows, horses, camels, goats, and sheep. But a few come from wild animals, such as lions, leopards, porcupines, deer, monkeys, rabbits, otters, and beavers. Lions are now extremely rare in Asia (there is still a small population in the Gir Forest of northwestern India), but some Asian mammals whose body parts are desired commodities for TCM are critically endangered, and it

* As long as there remain crippling and fatal diseases, there will be individuals willing to try alternatives to scientific treatment. Among today's "incurable" diseases are AIDS, arthritis, cardiovascular disease, some cancers, multiple sclerosis, and amyotrophic lateral sclerosis (ALS), also known as Lou Gehrig's disease. In many cases where Western medicine cannot provide a cure, patients turn to alternative solutions, which often take the form of herbal remedies. In the 1970s, laetrile, a substance derived from apricot pits, was heralded by some as a "natural" cure for cancer. Movie star Steve McQueen, diagnosed with a rare form of lung cancer, went to Mexico for laetrile treatments, which failed to save him, and the laetrile phenomenon subsided. His case is an example of people willing to try anything when they learn that the wonders of modern (Western) medicine have failed them. Die of cancer or try something—anything—that might save you?

is toward these—particularly tigers, rhinos, and bears—that later chapters are directed. This is in no way to be considered a blanket criticism of the principles or practices of traditional Chinese medicine, but rather it points out that some irresponsible people, often perverting the fundamentals of this venerable tradition, bear a large responsibility for the destruction of some increasingly endangered species. Of course there are millions of people in China, Southeast Asia, and elsewhere with little or no access to education on science who are ignorant of the composition of the potions they so eagerly consume and who know very little about the endangerment of animals. A wider understanding of the traditions, the medications, and the status of the endangered species might possibly save even more lives—human and animal.

3

Chinese Medicine, Western Medicine

When an American doctor named Benjamin Hobson (1808–71) arrived in Canton in 1851, he publicly deplored the sorry state of Chinese medicine as contrasted with Western "scientific" practice. Between 1851 and 1868 Hobson became a sort of medical missionary, trying to blend the two traditions by combining the spirituality of the Chinese with the science of the West. He published five textbooks, including one on gynecology and obstetrics, but his emphasis on surgical intervention rendered his procedures irrelevant to the practices of elite Chinese doctors, who totally rejected the idea of invasive surgery.

Almost a century later, in 1942, the young Mao Zedong was highly critical of Chinese medical practice as well. He wrote that "old doctors, circus entertainers, snake oil salesmen, and street hawkers are all of the same sort," a line that would have a truly devastating impact on traditional medicine twenty-five years later when Mao's words became the one and only source for the country's definition of political truth. This line was quoted in millions of copies of the red "Mao Bibles," serving as the Red Guard's main license for persecuting traditional Chinese medicine practitioners and criticizing their traditions. Mao nevertheless seems to have reversed himself during the period when he actually ruled China, and from 1954 to 1976, he reestablished traditional Chinese medicine, but combined it with Western influences. In some areas, Western medicine now appears to be supplanting traditional Chinese medicine, but the shift has been gradual, and many traditional practitioners remain. In a 1999 article in the *Journal of Chinese Medicine*,

Heiner Fruehauf, an authority on Chinese medicine, wrote of medical education at the bachelor's level:

> By far the most extensive classes are dedicated to Western medicine contents such as anatomy, physiology, immunology, parasitology, and other topics that are unrelated to the diagnostic and therapeutic procedures of classical Chinese medicine. From both a quantitative and a qualitative perspective, therefore, it would not be entirely inappropriate to state in slightly dramatized terms that the Chinese medicine portion in the contemporary TCM curriculum has been reduced to the status of a peripheral supplement—approximately 40% or less of the total amount of hours. . . . None of the specialty students, including acupuncture department graduates, are required anymore to familiarize themselves with the realm of original teachings, not even in the radically abridged form of classical quotations that still serve to bestow an air of legitimacy on most official TCM textbooks.

While Chinese medical practices now include elements of Western medicine, it is clear that a substantial number of courses will be devoted to traditional Chinese medicine (TCM), and there are still many practitioners who rely heavily on older, time-honored practices, such as acupuncture, herbal remedies, diet, exercise, and massage.

Traditional Chinese medicine is thousands of years older than its Western counterpart and founded on completely different principles. TCM looks at the bodily system as a whole; Western medicine looks at the structure and function of the parts. Western medicine manages disease; TCM works to maintain health. Western medicine is standardized; TCM is individualized. Western medicine is the result of laboratory experimentations; TCM is a summary of clinical observations. Western medicine mainly relies on medication and procedures; TCM emphasizes the role of the body in healing. Where TCM prescribes herbs and natural agents, practitioners of Western medicine emphasize chemical compounds—often derived from natural agents. While Western medicine is intended to be strictly science-based, TCM is considered a healing art. In the modern era of science and technology, it is not surprising that Western medicine has become the predominant system while TCM is regarded as an alternative or complementary form of healing. But

Western practitioners now recognize the validity of certain TCM practices and have incorporated them into established medial contexts.

Within Chinese cosmology, all of creation is a function of two polar principles, *yin* and *yang*: Earth and Heaven, winter and summer, night and day, cold and hot, wet and dry, inner and outer, body and mind. Harmony of these principles means health, good weather, and good fortune, whereas disharmony leads to bad luck, disease, and disaster. In Chinese medicine, the concepts of yin and yang are infinitely divisible but inseparable: one cannot exist without the other. In medicine, the concept of the necessary interdependence of yin and yang is used in explaining physiology, pathology, and treatment. Every person has a unique terrain to be mapped, a resilient yet sensitive ecology to be maintained. Just as a gardener uses irrigation and compost to grow robust plants, the doctor uses acupuncture and medicinals to recover and sustain health. Just as nature contains air, sea, and land, the human body is composed of *Qi* (pronounced "chee"), Moisture, and Blood. Qi is the animating force that gives humans the capacity to move, think, feel, and work. Moisture is the liquid medium that protects, nurtures, and lubricates tissue. Blood is the material foundation out of which bones, nerves, skin, muscles, and organs are created.

In TCM, the goal of treatment is to adjust and harmonize yin and yang—wet and dry, cold and heat, inner and outer, body and mind. This is achieved by regulating the Qi, in the Organ Networks: weak organs are strengthened, congested channels are opened, excess is dispersed, tightness is loosened, agitation is calmed, heat is cooled, cold is warmed, dryness is moistened, and dampness is drained. The duration of treatment depends on the nature of the complaint, its severity, and how long it has been present. Response varies; some need only a few sessions, whereas others need sustained care to reverse entrenched patterns established over time, practitioners say.

In origin, Chinese medicinals can be animal, vegetable, or mineral, in most cases simply prepared. Their properties are traditionally understood in terms of their Qi and flavor. The Qi of an agent is either warm/hot or cool/cold; warm and hot agents are used to treat cold patterns, and cool and cold agents are used to treat heat patterns. A neutral agent is one that is neither hot nor cold. The five flavors are acrid, sour, salty, bitter, and sweet; these flavors correspond to wood, metal, water, fire, and earth, and are often found to act upon the liver, lung, kidney, heart, and spleen, respectively. Medicinals are said to "enter" one or more channels. The substances affect the parts of the body through which the channels are believed to pass and can affect other agents in those regions. Functions of the medicinal agents are described in terms of restitution of aspects of the body (e.g., fortifying the spleen, supplementing the kidney) and in terms of eliminating evils (e.g., dispelling dampness, extinguishing wind).

Tiger bones, for example, are occasionally included in traditional Chinese prescriptions, but their uses are not particularly notable. Today, TCM does not actually equate tiger bones with sexual virility or shark's fin soup with ferocity. Here, for example, is the entire entry for tiger bone found in the section on "Wind-damp-dispelling medicinals" in the 1996 revision of Wiseman and Ellis's *Fundamentals of Traditional Chinese Medicine*:

> Warm, acrid, non-toxic. Enters the liver and kidney channels. Chases wind and settles pain; fortifies the bones and settles fright. Treats joint-running wind pain; hypertonicity of the limbs, limp lumbus and knees; fright palpitations; epilepsy; hemorrhoids and fistulas; prolapse of the rectum. *Directions*: Oral: decoct (9–15g), steep in wine, or use in pills or powders. *Caution*: Contraindicated in exuberant blood vacuity fire.

Nothing about enhanced virility. Evidently, the use of tiger penis as the "bone" creates a direct connection between the tiger's ferocity and the amorous capabilities of the consumer of the wine, pills, or powders. And because the cartilaginous fibers used in shark's fin soup are not a part of TCM at all—at least they have not appeared in any book I have consulted—any connection made between the soup and a man's performance is solely in the mind of the chef or the soup drinker.*

* In the West, when it was thought that sharks had some mysterious quality that prevented them from getting cancer, some enterprising entrepreneurs began killing sharks for their cartilage and marketing shark-cartilage pills as an anticarcinogen. They made a lot of money until a couple of researchers pointed out that sharks do get cancer—they even get cartilage cancer—and besides, taking shark-cartilage extract to prevent cancer was a little like eating the sawdust from redwood trees to make yourself taller.

In 1989, Zhang Enquin published the *English-Chinese Rare Materia Medica*, a handbook of herbal and animal drugs that are prescribed today in TCM, which he introduced in this way:

> Involving more than just the great contributions to the flourishing and prosperity of the Chinese nation, [the traditional medical pharmacy] represents an important chapter in the annals of Oriental civilization. Its unique theories and miraculous therapeutic effects have fascinated more and more people in the world. Included in this book are fifty clinically proved, valuable, and world-famous traditional Chinese drugs, of which twenty-seven are herbal drugs, twenty are animal drugs, and three are of other categories. [The three "other categories" are Chinese caterpillar, donkey-hide gelatin, and amber.]

To a Westerner, the use of animal parts as cures or preventatives may sound peculiar, but for people whose menu includes almost everything that can be raised or caught, animals such as sea horses, palm civets, raccoon dogs, and badgers as food or medicine are culturally quite acceptable.* David Kellogg, an American who moved to China in the early 1980s, wrote of his experiences in Hong Kong in his 1989 book, *In Search of China*:

> About a block from my hotel is the food market where you may buy roasted pangolins (a kind of scaly anteater that is really delicious), live Chinese raccoons, owls, parrot-like blue and green birds with straight beaks, monkeys, soft-shell turtles of all kinds, skinned, dried, and spiced rats, swans and geese, and, of course, dogs, usually split down the middle or hung by a hook through the throat.

---

* While many Westerners find badgers or civets unappetizing, they do not have a problem with animal parts *transplants*, technically known as xenotransplants from *xenos*, which means "strange" or "foreign" in Latin. In 1984, "Baby Fae" received a baboon heart and lived for three weeks before her immune system rejected it; baboon livers were given to two patients in 1992. Since then, the donor animal of choice has been the pig, and human patients have received pig liver transplants—usually a "bridge" until a human liver could be found—and experimental heart and kidney transplants have also been tried. Xenotransplant technology is still in the research stage, but if and when it is perfected, it will establish a further link between Western high technology and traditional Chinese use of animal parts in medicine.

The outbreak of SARS (Severe Acute Respiratory Syndrome) in Hong Kong, China, and Taiwan in the spring of 2003 brought to Western attention some of the more unusual food items on the Chinese menu, as it was suspected early on that the source might lie in the markets of southern China. On May 12, 2003, in response to the reports of SARS in palm civets, investigators from Animals Asia, a nonprofit conservation group based in Hong Kong, went through the Hua Nan Wild Animal Market in Guangzhou in southern China and found "cages and crate loads of masked palm civets, ferret badgers, barking deer, wild boars, hedgehogs, foxes, squirrels, bamboo rats, various species of snakes and endangered leopard cats, together with dogs, cats, rabbits and gerbils. As some of the traders attempted to hide their stash of wild animals, others insisted that theirs were captive bred—seemingly ignoring the fact that many animals showed bloody stumps, where their limbs had been severed in leg-hold traps in the wild."

From various sources, I have been able to find these additional animals eaten by the Chinese, but the list is hardly exhaustive; if an animal has four legs, wings, or even no legs at all, it can, it seems, find its way to the Asian dinner table: porcupines, silver foxes, badgers, squirrels, bamboo rats, common palm civets, spotted linsangs, wild boars, flying squirrels, flying foxes, mongooses, leopard cats, raccoon dogs, nutria, chipmunks, guinea pigs, horses, gerbils, donkeys, goats, chickens, peacocks, geese, ducks, quail, cobras, king cobras, rat snakes, water snakes, bamboo snakes, banded kraits, pythons, soft-shelled turtles, frogs, salamanders, and water monitors. The Chinese seem to have a special fondness for turtles; the Hong Kong and Guangzhou markets have special sections devoted to the display of various species, all of which are sold as food. In an Asian market in Cleveland, I saw live turtles and frogs offered as food items. In a market in Cambodia, writer Sy Montgomery asked a Cambodian biologist if there was any animal that people didn't eat there. "The vulture," he answered solemnly.

Some patterns of animal consumption may change in response to the threat of SARS. By the summer of 2003, civets, raccoon dogs, and many of the other exotic species that are staples of Guangdong's eclectic cuisine were gone from the markets, but it is unlikely that the ban will last beyond the SARS scare. Their disappearance doesn't mean the end of SARS;

indeed, it is not obvious that it originated in palm civets at all. The civets more than likely contracted the disease from other, more exotic species in the markets or some other place where they were in close contact. According to an article in *Science* for August 22, 2003, China lifted the four-month ban on selling masked palm civets and fifty-three other exotic species, because the Chinese government "did not find any evidence of a connection to SARS among those species." But then, in early 2004, when a man was diagnosed with SARS in Guangdong, the Chinese government decided that palm civets indeed harbored the SARS coronavirus and called for the destruction of ten thousand civets (Bradsher 2004).

Chinese eating habits, developed over thousands of years, are not based on the popularity or "cuteness" of the animal or what others may regard as exotic. As a matter of fact, neither are Western eating habits: those who would condemn the eating of badgers or porcupines should remember that Americans and Europeans have always hunted and eaten deer, wild sheep, rabbits, ducks, and other game birds such as pheasants and grouse; in medieval Europe, songbirds—like the "four and twenty blackbirds" baked in Old King Cole's pie—were regularly consumed. "White hunters" in Africa traditionally shoot and eat various antelopes, zebras, and other game animals when on safari.

A wide variety of vertebrate species are used in commercial trade in China's cities, especially in south China, Li Tining and David Wilcove (2004) point out, and they estimate that over 1,500 animal species are used in TCM. The consumption of wild animals is not analogous to traditional Chinese medicine, however, although both have been practiced for ages. TCM is three thousand years of carefully researched and tested practices. Though some of these may understandably appear strange to Western eyes, many of the innovations usually ascribed to Western physicians or medical researchers may actually have occurred in ancient China. For example, most Westerners believe that William Harvey discovered the circulation of the blood. But in *The Genius of China*, Robert Temple (1998) commented, "Harvey was, however, not even the first European to recognize the concept, and the Chinese had made the discovery two thousand years before." Temple maintained that "the ancient Chinese conceived of two separate circulations of fluids in the body.

Blood, pumped by the heart, flowed through the arteries, veins, and capillaries. *Ch'i*, an ethereal, rarified form of energy, was pumped by the lungs to circulate through the body in invisible tracts. The concept of this dual circulation of fluids was central to the practice of acupuncture."

The view that Chinese medicine understood the basics of heart circulation long before Western medicine is still contentious. In *Medicine: An Illustrated History*, Albert Lyons wrote, "Ideas in the *Nei Ching* concerning movement of the blood ('All the blood is under control of the heart'; 'The blood flows continuously in a circle and never stops'), have been thought to approach an understanding of its circulation antedating Harvey by thousands of years; however, some body vessels were thought to convey air, and there is little evidence that commentators perceived the blood-carrying vessels as a contained system." Temple, on the other hand, believes that the Chinese fully understood the system, drawing on this comment in *Nei Ching*: "What we call the vascular system is like dykes and retaining walls forming a circle of tunnels which control the path that is traversed by the blood so it cannot escape or find anywhere to leak away."*

Early Chinese Medicine

The *Huang-ti Nei Ching* (The Yellow Emperor's Classic of Internal Medicine) is generally considered the first discussion of Chinese medicine. It is reputed to have been compiled by Huang-ti, the "Yellow Emperor," around 2,600 BC, transmitted orally for centuries, and finally committed to writing around the third century AD. Joseph Needham (1900–95), the foremost Western authority on the history of science in China and author of a detailed history of Chinese medicine, believed there really was no

* William Harvey (1578–1657) was fluent in Greek and Latin, but not Chinese, so he would have had no access to the earlier Chinese studies. He was instead originally influenced by Galen's idea that there were two types of blood, venous and arterial, that followed different pathways and served different functions. In his 1628 *Anatomical Essay Concerning the Movement of the Heart and the Blood in Animals*, he deduced that there was only one circulatory system and that the blood was circulated through the heart muscle by the ventricles, and not absorbed and replenished by the liver, as Galen had suggested.

Yellow Emperor. Similarly, in the introduction to her translation of the *Huang-ti Nei Ching*, Ilza Veith suggests that "modern historiography tends to relegate him to the realm of legend." Controversy surrounds estimates of *when* the *Huang-ti Nei Ching* was written. Some authorities cling to the 2697 BC date, but research by Chinese medicine scholar Wang Chi-min suggests that it was composed around 1000 BC, which would make its author (or authors) contemporaries of Hippocrates. Veith again:

> If the history of the *Yellow Emperor's Classic* is to be compared with that of the *Corpus Hippocraticum*, which originated at about the same time, a curious and somewhat contradictory development may be noted. The works of the Greek tradition were composed to serve as text-books for the practitioner, yet the practical value of their contents was superseded centuries ago. Apart from their significance for the medical historian, the value of these works has for centuries consisted in creating for the Western physician the moral and ethical concept of the ideal physician. On the other hand . . . China's earliest book concerned with the art of healing was never meant to be a mere text-book of medicine, but rather a treatise on the philosophy of health and disease; and yet it was taken over by the physician, not as a guide towards an ideal life, but as a help for the actual practice of medicine.

However and whenever it was written, the *Nei Ching* consists of questions asked by the (legendary) emperor with long answers provided by Ch'i Po, his (also legendary) prime minister. All phases of health and illness are discussed, including prevention and treatment, ethics, and daily regimens, incorporated into an inclusive system that combines the Tao, yin and yang, and the theory of the five elements (Metal, Water, Wood, Earth, and Fire). Veith summarizes:

> Man, according to the system propounded in the *Nei Ching*, was subdivided into a lower region, a middle region, and an upper region, and each of these regions was subdivided three times, each subdivision containing an element of heaven and an element of man. This scheme of subdivisions was concurrent with another

scheme according to which each of the three main subdivisions was held to be composed of one part Yin and one part of Yang; i.e., the human body was regarded as consisting of three parts of Yin and three parts of Yang. Since treatment of a specific disease or a specific organ depended largely on its location within a particular part of Yin or Yang, knowledge of these subdivisions was very important for diagnosis as well as for treatment.

It was in the *Nei Ching* that the seats of the particular "spiritual resources" were identified: happiness dwells in the heart, thought and ideas in the spleen, "inferior spirit" (sorrow) in the lungs, the will controlled in the kidneys, and the liver houses anger as well as the soul. The practice of Chinese medicine consisted of diagnosing illness—primarily by a meticulous analysis of the pulse—and attempting to rectify the imbalances in the five spiritual resources; the five climates (heat, cold, wind, humidity, dryness); the five viscera (heart, lungs, liver, spleen, kidneys); and, as we've seen, the five flavors (salty, bitter, pungent, sour, sweet) by rebalancing the perceived discrepancies. Many other factors were considered in the diagnosis, including the direction of the wind, the season of the year, the color of the patient's skin, and even the patient's dreams.

The four diagnostic procedures a practitioner would use, according to Needham (2000), are physical inspection of the patient, listening to the sounds of the body (auscultation) through a stethoscope or similar instrument, taking the patient's medical history, and examining the body by touch (palpation). Except in marriage, contact between the sexes was prohibited, and doctors, all of whom were men, were forbidden to examine women. The female patient would often extend her arm through the bed curtains for the doctor to take the pulse, which was by far the most important element in Chinese diagnosis. (In some instances, the doctor carried an ivory figure of a woman so that the patient could point to the area where the pain was felt.)

The Chinese philosopher Confucius (actually *K'ung Fu-tse*), who lived some five hundred years before the birth of Christ, forbade violation of the human body, so many Chinese medical practices relied upon reasoning and assumption rather than dissection or even direct observations.

"It was not entirely the superiority of Chinese internal medicine that made surgery unnecessary," Veith comments, "but the Confucian tenets of the sacredness of the body, which counteracted any tendency toward the development of anatomical studies and the practice of surgery."

The use of needles was not considered a violation of the body, and acupuncture and what was called *moxibustion* were held to be the primary forms of applied treatment, but neither is explained in the *Nei Ching*, suggesting that these practices existed even before the appearance of the Yellow Emperor's book or didn't come along until later. As still practiced today, acupuncture consists of inserting sharp needles of various sizes into particular points of the body along twelve channels or "meridians" that occur in pairs arranged symmetrically on the left and right sides of the body and are believed to be related to the various organs. The channels are deeply embedded in the muscles but come to the surface at 365 points that are available for "needling." The needles are inserted with a twirling motion, and the depth, angle, and duration of the insertion are left to the judgment of the acupuncturist. Sometimes acupuncture is combined with moxibustion, which consists of holding a stick of burning moxa (a variety of mugwort, *Artemisia vulgaris*) over the needled area, or when applied alone, placing on the skin powdered leaves of moxa, which are then ignited. This can be effected with the leaves directly on the skin or in a closed capsule that is heated and applied to the affected area.

Perhaps the most important idea contained in Chinese medicine is the one that suggests that prevention is more important than cure. In the *Nei Ching* we read: "To cure disease is like waiting until one is thirsty before digging a well, or to fabricate weapons after the war has commenced." Diet was also an important component in maintaining the body's balance, and chapter 22 of the *Nei Ching* ("Treatise on the Seasons as Patterns of the Viscera") specifies that the five flavors—pungent, sour, sweet, bitter, salty—have softening, dispersing, gathering, retarding, and strengthening effects, respectively. An example:

> Those who suffer from a disease of the kidneys are quick-witted and active at midnight, and their spirits are heightened during the entire days of the last months of Spring, Summer, Fall, and Winter, and

> they become calm and quiet toward sunset. Sick kidneys have the tendency to harden; then one should eat bitter food to strengthen them. One uses bitter food to supplement and to strengthen them and one uses salty food to drain them and to make them expel.

Along with acupuncture, the pharmacopoeia served as the foundation of Chinese medicine, and many of the precepts delineated in the ancient *Huang-ti Nei Ching* are in practice today, not only in China, but around the world. Interestingly, there is nothing in the *Nei Ching* about the use of herbs or animal parts as medicine, and indeed, there is nothing in the Yellow Emperor's book that discusses the ingestion of anything other than known food substances to correct imbalances. Centuries would pass before the subject of using herbs and animal substances for medicinal purposes was published, but it is likely that it had been going on long before it was encoded in books.

Sometime during the Han Dynasty (206 BC to AD 220), the *Divine Husbandman's Classic of the Materia Medica* (*Shen nong ben cao jing*) was published, the first Chinese text to focus on the medicinal use of various substances and not only the atmospheric, anatomical, and philosophical dialogues of the *Nei Ching*. The *Divine Husbandman's Classic* contains 364 entries, one for each day of the year, 252 of which were botanical, 45 mineral, and 67 zoological. It would be another one thousand years before the *Shen Nung Materia Medica* was written, detailing the pharmacological use of three hundred substances, including antipyretics, cathartics, diuretics, emetics, sedatives, stimulants, digestive remedies, antidiarrheal medicaments, and mercury and sulfur for skin diseases.

The Imperial Institute of Physicians, set up by the Tang Dynasty emperors in the seventh century AD, was the world's first medical school. It had an enrollment of some 350 students, specializing in medicine, surgery, or acupuncture—then considered the three divisions of traditional medicine. By this time, invasion of the body was no longer prohibited, as the medical profession recognized that for skin diseases, hemorrhoids, and the treatment of fractures, wounds, and septic conditions, some part of the body might have to be "violated." China's first hospital was established in AD 510 to cope with an epidemic in Shansi

Province, and during the following centuries a number of government-organized hospitals for lepers and the poor were set up.

In 1597, after twenty-seven years of research, the great pharmacologist Li Shih-chen published the *Pen Ts'ao Kang Mu* (Compendium of Materia Medica), which listed 1,892 drugs and some ten thousand prescriptions. One thousand of the drugs were of vegetable origin, four hundred were zoological, and the remainder were mineral. The *Pen Ts'ao Kang Mu* contains more than five hundred suggestions to strengthen and maintain the body, many of which were evidently developed by Li Shih-chen himself. Among the "zoological" drugs were pig's epiglottis, buffalo's nose, porcupine's urine, the meat of animals killed by thunder, and just about every part of a tiger.

The Origins of Western Medicine

At the outset, early Chinese and early Western medicine were not that different; for the most part, the origins of diseases were a mystery, and we have no way of knowing how effective either system actually was. In the West, clinically similar to the teachings of the *Nei Ching* were those of Hippocrates, the Greek physician and scholar born around 460 BC on the island of Kos near the western coast of Asia Minor. His writings—collectively known as the "Hippocratic Corpus"—have come down to us in the form of sixty books, originally housed in the great Library of Alexandria, but copied and rewritten countless times. It is unlikely that the *Corpus Hippocraticum*, which includes sections on anatomy, physiology, pathology, therapy, diagnosis, prognosis, surgery, gynecology and obstetrics, mental illness, and ethics, was written by one man, however. The books of the *Corpus*, the social historian of medicine Roy Porter (1997) tells us, "derive from a variety of hands, and, as with the books of the Bible, they became jumbled up, fragmented, and then pasted together in antiquity. . . . Scholarly ink galore has been spilt as to which were authentic and which spurious; the controversy is futile."

Hippocrates believed that disease resulted from an imbalance of the four bodily humors and that the goal of medicine was to restore the balance through appropriate diet and hygiene, turning to more drastic treatment only as a last resort. As Sherwin Nuland wrote in *Doctors*:

> The Hippocratic physicians saw diseases as events that happen within the context of the life of the entire patient, and they oriented their treatment toward restoration of the natural conditions and defenses of the sick person and the re-establishment of his proper relationship to his surroundings. . . . It was the basically holistic clinical approach of Hippocrates that provided the clear light which led Greek medicine out of the mire of theurgy and witchcraft.

Hippocrates taught that nature seeks an equilibrium of the four "humors"—blood, yellow bile, black bile, and phlegm—which were constantly renewed by the food one ate. These humors were mixed and moved in the body by "innate heat," which was a form of energy generated by the heart and the essential ingredient in human composition. Furthermore, the four humors corresponded to the four elements—fire, air, earth, and water—which represent the qualities of heat, dryness, cold, and dampness. (This, of course, was not very different from the precepts of early Chinese medicine, even though the details were different.) The Greeks, with few available pharmacological remedies, believed in diagnosing the condition of the patient and, where necessary, prescribing such things as purgatives, emetics, baths, bloodletting, wine, bland drinks, and a calm atmosphere, all designed, as Nuland points out, to aid nature in its attempts to rid the body of excessive humors.

Despite his elaborated view of bodily processes, Hippocrates recognized that there was more unknown than known about healing. His famous aphorism "Art is long, but life is short" (*Ars longa, vita brevis*) is now applied to any and all arts, but Hippocrates meant that the art of healing has a much longer life than that of its practitioners.* Of course, what he is best known for is the vow that bears his name. The Hippocratic

* The complete statement is "Life is short, and art is long; opportunity fleeting, experience delusive, judgment difficult, and the crisis grievous. It is necessary for the physician not only to provide the needed treatment but to provide for the patient himself, and for those beside him, and to provide for his outside affairs."

oath, sworn to even today by graduating medical students, as Albert Lyons concisely put it in a 1987 book, "contains both affirmations and prohibitions. It begins with pledges to the gods and to teachers as well as future students. The prohibitions are against harm to the patient, deadly drugs, abortion, surgery, sexual congress with the patient or his household, and revelation of secrets discovered while ministering to the sick. The duties are to act with purity and holiness."

Although the oath endures, the most influential of the ancient doctors was not Hippocrates but Galen, a Greek born at Pergamum in AD 129, during the reign of the Roman emperor Hadrian. In his teens, Galen became a *therapeutes*, an attendant upon the healing god Asclepius, and upon the death of his father, he inherited enough money to allow him to travel and broaden his medical horizons. When Galen returned home, he was named to the prestigious post of physician to the gladiators, and he learned much about the human body from his examination of gladiators' wounds. In 168, he left Pergamum for Rome, where he continued to write, lecture, and practice medicine, numbering the emperors Marcus Aurelius and Commodus among his patients. Unlike the works ascribed to Hippocrates, those of Galen seem to have been written by him, and his place in the history of medicine is assured by the realization that, as Porter wrote, "he was an erudite man and an accomplished philosopher, particularly in constructing an image of the organism as a teleological unity open to reasoning. For him, anatomy proved the truth of Plato's tripartite soul, with its seats in the brain, heart and liver; and Aristotelian physics with its elements and qualities explained the body system."

Galen's views were to dominate Western medicine for 1,500 years. In addition to summarizing the state of medicine at the height of the Roman Empire, Galen reported his own important advances in anatomy, physiology, and therapeutics. He made a special study of the pulse and showed that arteries carried blood, not air as many believed, and he made important discoveries about the spinal cord and nervous system that would not be appreciated until the nineteenth century. He elaborated on the four humors by classifying all personalities into four types: choleric, sanguine, phlegmatic, and melancholic—terms still in use today to characterize dispositions if not physiological conditions. He advocated large-scale use of medications and often prepared his own prescriptions, mixing agents whose properties he identified as hot, cold, dry, or moist. (Lyons gives an example of one of his recommendations, in which "an illness categorized as hot required a drug that was in the cold category"—a view also at the root of Chinese prescriptive medicine.) One of Galen's favorite remedies was theriac—the word comes from the Greek *therion*, which means "beast"—a potpourri of ingredients he prescribed to combat everything from inflammations to poisons and pestilence. One of the ingredients was probably the flesh of poisonous snakes. Like the Chinese, Galen believed in the use of animal parts as well as herbs in treating illness.

By the sixth century, many of Galen's works had been translated into Latin, and with the rise of Arab-Muslim power in the Mediterranean, they were translated into Arabic as well. Muslim science was a repository upon which Western societies drew again and again. Physicians such as Rhazes (born in Persia), Avicenna, Albucasis, Averroës, and Maimonides (all Spanish-born) devised new treatments but employed many of the same criteria for diagnosing an illness as Hippocrates did: behavior, excreta, bodily effluvia, swellings, and the location and character of pain. Islamic medicine favored cauterization for internal and external diseases and prescribed drugs of all kinds, many of which—for example, nutmeg, ambergris, camphor, cloves, tamarind, myrrh, and senna—had to be imported from India or China. "The value of Arab contributions to medicine," wrote Porter, "lies not in their novelty but in the thoroughness with which they preserved and systematized existing knowledge."

It was not until the Muslim invasions of Africa, Spain, and parts of France that medicine replaced prayer as a means of curing the sick in Europe. Throughout the Middle Ages, during which the Europeans seemingly avoided any reminiscences of the ancient world, they prayed to God and to assorted saints, who sometimes let them down. Despite their prayers, smallpox was rampant throughout Europe, and in 1347, the bubonic plague, or Black Death, killed at least one-quarter of the population of Europe. "In the earlier Middle Ages, abbeys and monasteries were the repositories of medical knowledge," the historians Joseph and Francis Gies tell us in *Life in a Medieval City*. "The principal effect of their regime was to repeal Hippocrates' law that illness is a natural phenomenon and to make it appear to be a punishment from on high." And Robert Lacey and Daniel Danziger, in their 1999 study of first-millennium England, commented:

> The sign of the cross was the antiseptic of the year 1000. The person who dropped his food on the floor knew that he was taking some sort of risk when he picked it up and put it in his mouth, but he trusted in his faith. Today we have faith in modern medicine, though few of us can claim much personal knowledge of how it actually works, and we also know that the ability to combat quite major illnesses can be affected by what we call "a positive state of mind"—what the Middle Ages experienced as "faith."

Belief in divine intervention, however, did not stop the Anglo-Saxons from applying potions of all sorts to afflicted unfortunates. Citing a tenth-century Winchester document known as "Bald's Leechbook," Lacey and Danziger provide a vivid example:

> Bald's prescription for dysentery showed a particularly well-balanced combination of folk remedy, religious conviction, and tender loving care—which probably constituted the most efficacious ingredient in the recipe: "Take a bramble of which both ends are in the earth, take the newer root, dig it up, and cut nine chips on your left hand, then sing three times: *Miserere mei dens* [Psalm 56] and nine times the Our Father. Take then mugwort and everlasting and boil these three in several kinds of milk until they become red. Let him then sup a good bowl full of it, fasting at night, sometime before he takes other food. Make him rest in a soft bed and wrap him up warm.

The leading pharmacological text for sixteen centuries and the foremost classical source of modern botanical terminology was *De materia medica*, the work of Dioscorides (c. AD 40–90), a first-century Greek physician and pharmacologist. Dioscorides' travels as a surgeon with the armies of the Roman emperor Nero provided him an opportunity to study the features, distribution, and medicinal properties of many plants and minerals. Written in five volumes around the year 77, *De materia medica* contains excellent descriptions of nearly six hundred plants, including cannabis, colchicum, water hemlock, and peppermint, and includes descriptions of approximately one thousand simple drugs.

In medieval Europe, accumulated experiences of the curative powers of various plants were often codified into books known as *herbals*. Along with painstakingly accurate illustrations of various botanicals were found descriptions of their medicinal or magical properties much like their Chinese counterparts. Often incorporated into these herbals was the "doctrine of signatures," based on the resemblance of certain plants or plant parts to specific human organs or parts. Thus, heart-shaped leaves were thought to relieve heart disease; the convoluted walnut resembled the human brain and was therefore employed to relieve brain disorders; and the deep-throated figworts were recommended for scrofula, the swelling of the glands in the throat. Mandrake (*Atropa mandragora*), a member of the nightshade family, which includes belladonna, henbane, and tobacco, however, was the plant most infused with magical properties. The long root, which can sometimes resemble a human form, has been used since ancient times to arouse ardor, overcome infertility, and even increase wealth. It is poisonous, a narcotic, an anesthetic, and a preventative against demonic possession. It was reputed to grow only under the gallows of murderers. It screamed like a human when pulled from the ground, and whoever heard it was killed or driven mad. The only way to pull it out of the ground was to tie a dog to it; the dog would die, but at least you had the root. In *Romeo and Juliet*, Shakespeare's line is, "And shrieks the mandrake torn out of the earth, that living mortals hearing them run mad," and the Elizabethan poet John Donne (1571–1631) wrote:

> Go, and catch a falling star,
> Get with a child a mandrake root,
> Tell me, where all past years are,
> Or who cleft the Devil's foot.

Because magic, religion, and philosophy as well as the scientific disciplines of biology, medicine, and botany all originally coexisted and evolved together over time, each contributed to the elaborate rituals that developed around the gathering and use of plants and herbs. How they were used was important too. To fend off demons or cure diseases, herb drinks were mixed with ale, milk, or vinegar; many of the potions were made with herbs mixed with honey. Ointments concocted with herbs and butter were prescribed for common ailments such as bleeding noses, baldness, sunburn, loss of appetite, and dog bites. Herbs were also utilized as amulets or charms against evil and diseases. One might hang them from the door (usually with red wool), to preserve one's eyesight, cure lunacy, prevent one from fatigue while traveling, or even protect cattle.

One of the most important herbals of the Elizabethan era was *Historie of Plants* by John Gerard (1545–1612). For his botanical descriptions and remedies, Gerard depended on his own observations, but he also consulted earlier authorities. Of the plant he called "Solomon's Seale" (*Polygonatum multiflorum*), for example, which produces clusters of greenish white, tubular flowers and dark, shiny berries in the fall, he wrote:

> Especially among the vulgar sort of people in Hampshire, *Galen*, *Dioscorides*, or any other have not so much as dreamed of; which is, that, if any of what sex or age soever chance to have any bones broken, in what part of their bodies soever, their refuge is to stampe the roots hereof and give it unto the patient in ale to drinke, which sodoreth and glues together the bones in very short space, and very strangely, yea, although the bones be but slenderly and unhandsomely placed and wrapped up. . . . The root stamped and applied in the manner of a pultesse, and laid upon members that that have beene out of joynt, and newly restored to their places, driveth away the paine, and knitteth the joynt very firmely, and taketh away the inflammation if there chance to be any.

Artfully blending herbalism, alchemy, and astrology, Nicholas Culpeper (1616–54) gave us an insightful, often amusing glimpse of European medicine in the seventeenth century. His dependence on his predecessors can be seen in his admonition: "I shall desire thee, whoever thou art, that intendest the noble (though too much abused) study of physic, to mind heedfully these following rules; which being well understood, shew thee the Key of *Galen* and *Hippocrates* their method of physic: he that useth their method, and is not heedful of these rules, may soon cure one disease, and cause another more desperate." In a potent potpourri of cures, Culpeper suggests defenses not only against disease and infirmity, but against witchcraft, lustfulness, melancholy, intemperate dreams, vipers, serpents, mad dogs, and the plague.

Like those in medieval China, most Western herbals emphasized botanical preparations. Culpeper's 1653 *The Complete Herbal and English Physician Enlarged* is predominantly botanical, but it also contains a section on the use of "Living Creatures," such as millipedes, scorpions, earthworms, and ants; and another section headed "Parts of Living Creatures, and Excrements," which are prescribed for specific ailments. For example, "The brain of *Sparrows* being eaten, provokes lust exceedingly," and "A flayed *Mouse*, dried and beaten into a powder, and given at a time, helps such as cannot hold their water, or have a *Diabetes*, if you do the like three days together." Even a rare commodity such as *Ivory* or *Elephant's tooth*, "binds, stops the *Whites*, it strengthens the heart and stomach, helps the yellow jaundice, and makes women fruitful." *Whey*, a coagulation of milk still used in cheese-making, can be used to cure as many ailments as any Chinese medicament, as it "attenuates and cleanses both choler and melancholy; wonderfully helps melancholy and madness coming from it; opens stoppings of the bowels; helps such as have the dropsy and are troubled with stoppings of the spleen; rickets and hypochondriac melancholy: for such diseases you may make up your physic with whey. Outwardly it cleanses the skin of such deformities as come through choler or melancholy, as scabs, itch, morphew, leprosies, &c."

Lest you assume that it was only Chinese pharmacies that stocked animal parts, here is Culpeper's catalogue of the "Parts of Living Creatures and Excrements" that British apothecaries had to keep on hand:

> The fat, grease, or suet of a Duck, Goose, Eel, Boar, Herron, Thymallows (if you know where to get it), Dog, Capon, Beaver, wild Cat, Stork, Coney, Horse, Hedgehog, Hen, Max, Lion, Hare, Pike, or Jack (if they have any fat, I am persuaded 'tis worth twelve-pence a grain), Wolf, Mouse of the mountains (if you can catch them), Pardal, Hog, Serpent, Badger, Grey or brock, Fox, Vulture (if you can catch them), Album Graecum, Anglice, Dog's dung, the hucklebone of a Hare and

> a Hog, East and West Bezoar, Butter not salted and salted, stone taken out of a man's bladder, Vipers flesh, fresh Cheese, Castorium, white, yellow, and Virgin's Wax, the brain of Hares and Sparrows, Crabs' Claws, the Rennet of a Lamb, a Kid, a Hare, a Calf, and a Horse, the heart of a Bullock, a Stag, Hog, and a Wether, the horn of an Elk, a Hart, a Rhinoceros, an Unicorn, the skull of a man killed by a violent death, a Cockscomb, the tooth of a Bore, an Elephant, and a Sea horse, Ivory, or Elephant's Tooth, the skin a Snake hath cast off, the gall of a Hawk, Bullock, a she Goat, a Hare, a Kite, a Hog, a Bull, a Bear, the cases of Silk-worms, the liver of a Wolf, an Otter, a Frog, Isinglass, the guts of a Wolf and a Fox, the milk of a she Ass, a she Goat, a Woman, an Ewe, a Heifer, the stone in the head of a Crab, and a Perch, if there be any stone in an Ox Gall, the Jaw of a Pike or Jack, Pearls, the marrow of the Leg of a Sheep, Ox, Goat, Stag, Calf, common and virgin Honey, Musk, Mummy, a Swallow's nest, Crabs Eyes, the Omentum or call of a Lamb, Ram, Wither, Calf, the whites, yolks, and shells of Hen's Eggs, Emmet's Eggs, bone of a Stag's heart, an Ox leg, Ossepir, the inner skin of a Hen's Gizzard, the wool of Hares, the feathers of Partridges, that which Bees make at the entrance of the hive, the pizzle of a Stag, of a Bull, Fox Lungs, fasting spittle, the blood of a Pigeon, of a Cat, of a he Goat, of a Hare, of a Partridge, of a Sow, of a Bull, of a Badger, of a Snail, Silk, Whey, the suet of a Bullock, of a Stag, of a he Goat, of a Sheep, of a Heifer, Spermaceti, a Bullock's spleen, the skin a Snake hath cast off, the excrements of a Goose, of a Dog, of a Goat, of Pigeons, of a stone Horse, of a Hex, of Swallows, of a Hog, of a Heifer, the ankle of a Hare, of a Sow, Cobwebs, Water thells, as Blatta Baxantia, Buccinae, Crabs, Cockles, Dentalis, Entails, Mother of Pearl, Mytuli Purpurae, Os sepiae, Umbilious Marinas, the testicles of a Horse, a Cock, the hoof of an Elk, of an Ass, of a Bullock, of a Horse, of a Lyon, the urine of a Boar, of a she Goat.

Most of these animal parts, while they sound exotic to us, were not uncommon in medieval England (well, maybe the rhinoceros, elephant, and unicorn were not so easily found in British farmyards), and the parts could be collected from animals that died of natural causes or were butchered for food. By and large, Culpeper's animal pharmaceuticals did not endanger animal species. (Neither, at the time, did those of the Chinese, given the scale of medications prescribed and the number of animals that would have to be killed to provide them.)

The second half of Culpeper's *Complete Herbal and English Physician Enlarged* is described by its author as "an astrologo-physical discourse of the common herbs of the nation; containing a complete Method or Practice of Physic, whereby a Man may preserve his Body in Health, or cure himself when sick, with such things only as grow in England, they being most fit for English Constitutions." We find directions for making familiar concoctions such as syrups, conserves, oils, plasters, poultices, and pills, but also more arcane applications such as juleps, decoctions, electuaries, lohochs, and troches. Culpeper's purpose in delineating all these cures is "to preserve in soundness and vigour, the mind and understanding of man; to strengthen the brain, preserve the body in health, to teach a man to be an able co-artificer, or helper of nature, to withstand and expel Diseases." Just as in the Chinese herbals, the *English Physician Enlarged* identifies certain conditions that can be treated or cured with specific preparations. For example, "to stop fluxes of the blood, the menses, the haemorrhoids or piles, [and] also help ulcers in the breast or lungs," you must make a troche (lozenge) thus:

> Take of Amber an ounce, Hart's-horn burnt, Gum arabic burnt, red Coral burnt, Tragacanth, Acacia, Hypocistis, Balaustines, Mastich, Gum Lacca washed, black Poppy seeds roasted, of each two drams and two scruples, Frankincense, Saffron, Opium, of each two drams, with a sufficient quantity of mussilage of the seeds of Fleawort drawn in Plantain Water, make them into troches according to the art.

While both English and Chinese apothecaries used a surprising abundance of animal parts (and excrements), it was still the botanicals that dominated both materiae medicae. Of course, the plants differed according to what was available. Where the English materia medica included bracken, cowslip, elm, heather, lavender, and woad, the Chinese version includes ginseng, lotus root, mung bean, sandalwood, cinnamon, and gardenia. Regardless of the geographical differentiation, however, the English herbals often read as if the cures were lifted from a book of Chinese medicine, indicating a surprising overlap of the two systems, if not as often in their specific claims. Here, for example, are the medieval medicinal uses of common wormwood (*Artemesia absinthium*) from *Brother Cadfael's Herb Garden*, Rob Talbot and Robin Whiteman's authoritative study of medieval plants and their uses: "Prescribed for stimulating the appetite, aiding digestion, treating jaundice, constipation and kidney disorders, reducing fevers, expelling worms from the intestine, remedying liver and gall bladder complaints, curing flatulence and improving blood circulation." And now the uses of sweet wormwood (*Artemesia* sp.) from Wiseman and Ellis's *Fundamentals of Traditional Chinese Medicine*: "Clears heat, resolves summerheat . . . treats warm disease, fever, malaria, dysentery, jaundice, scab, and itching."

While the practice of herbals and faith healing were in full flower, another strand was developing, which would ultimately come to dominate curative practices in the West and for which there was no real corollary in China at the time. Galen was not translated into Latin until 1476. The medical faculty of the University of Paris adopted his works as their standard text and stuck to his words as if they were gospel. But the man known as Paracelsus soon completely changed the way medicine was taught in Europe. Born in Switzerland around 1493 as Theophrastus Philippus Aureolus Theophrastus Bombast von Hohenheim, Paracelsus was well-versed in alchemy, chemistry, and metallurgy, but his fame lies in his boisterous and argumentative rejection of traditional theories of medicine. Teaching in German instead of Latin and wearing a leather apron instead of academic robes, Paracelsus refuted the teachings of Galen and Hippocrates, and ridiculed Galen's humoral theory of disease, advocating instead the use of specific remedies for specific diseases, many of which included chemicals such as laudanum, mercury, sulfur, iron, silver, gold, and arsenic. In their summary of Western medicine in *Chinese Herbal Medicine*, Bensky and Gamble accuse Paracelsus (whom they dub "a magician of the Renaissance") of being responsible for the decline of herbalism in the West, because he felt "that the action of a remedy did not depend on its hypothetical qualities (such as hot or cold) but rather on its specific healing virtue. Herbs were too imprecise or crude. . . .

There was a powerful drug for each disease just waiting to be discovered." Many of his remedies employed the popular doctrine of signatures, in which a plant that resembled an organ was used to cure a disease, such as the plant known as "eyebright" (*Euphrasia officinalis*), which was used to treat eye problems. He rejected dissection of cadavers as providing no information on how living systems actually worked, and although he died in 1541, "before Vesalius published his *Fabrica* . . . he would probably have deemed it not worth a sausage" (Porter 1997).

If Paracelsus had not led the way, Andreas Vesalius (1514–64), professor of medicine at the University of Padua, would never have overthrown Galen's dogma. Vesalius' emphasis on dissection and anatomy radically changed the direction of medical examination and pushed Western medicine further from the precepts of traditional Chinese practices. At Paris, Vesalius studied medicine in the Galenic tradition, and although he acquired great skill in dissection, he remained under the influence of the master's misguided concepts of human anatomy. But where his predecessors had watched from on high while a barber-surgeon pulled organs out of a cadaver, Vesalius performed his own dissections and eventually produced four large anatomical charts. In 1539, he produced *Institutiones anatomicae*, an anatomical manual for his students, in which he began to question some of the Galenic precepts. By 1540, he was certain that Galen's research described the anatomy of an ape and not of a human, and he argued that Galen's errors could only be repudiated by active dissection and observation of the human corpus. In 1543, the twenty-nine-year-old Vesalius produced what Joseph Petrucelli (1978) called "one of the greatest books in the history of man," *De Humanis Corporis Fabrica* (Structure of the Human Body), with illustrations by Vesalius himself and students from Titian's studio, in which the discrepancies between Galen's descriptions (of an ape) and his own (of a human) were delineated.

Was there cross-fertilization between Western and Chinese medicine before the modern age? Did Marco Polo bring back news of Chinese medicine when he returned to Europe in 1295? Probably not (there is no mention of doctors or medicine in his *Travels*), but about a century after Polo's accounts were published, the Chinese themselves decided to explore the world outside their kingdom. Under the leadership of admiral Cheng Ho (1371–1433), fleets were sent far and wide on voyages of exploration, trade, and tribute collecting. Known as the "Three-Jewel Eunuch," Cheng Ho was born a Muslim in Yunnan Province, captured when he was ten, and brought to the court of the newly established Ming dynasty, where he was castrated and sent into the army as an orderly. Eventually a confidant and advisor to Zhu Di, the emperor who built the Forbidden City in Peking, he was appointed commander of missions to the "Western Oceans." From 1405 to 1433, fleets of Chinese armed warships, some more than 400 feet long, made seven epic voyages through the seas of China and the Indian Ocean. They carried cargoes of silks, porcelains, and lacquerware to trade for ivory, pearls, and spices, but their main purpose was to impress local rulers with the riches of the Chinese empire and the grandeur of its emperor. The expeditions visited such far-flung ports of call as Sumatra, Calicut, Ceylon, Siam, Malacca, the Maldives, the Ryukyus, Brunei, and Mombasa and Malindi on the east coast of Africa. The seventh expedition (1431–33) was the most ambitious of all, carrying forty thousand men to every port from Java to Mecca and returning with tributes collected from the Asian and Arab states, including horses, elephants, and a giraffe, but without Cheng Ho, who had died at sea.

Zhu Di's successors had no interest in replicating his grandiose naval demonstrations, however. They forbade all overseas travel, curtailed foreign trade, shunned contact with other nations, and disbanded the fleet, along with China's policies of outward expansion. As Louise Levathes wrote in her 1994 *When China Ruled the Seas*:

> The period of China's greatest outward expansion was followed by the period of its greatest isolation. And the world leader in science and technology in the early fifteenth century was soon left at the doorstep of history, as burgeoning international trade and the beginning of the Industrial Revolution propelled the Western world into the modern age.

By the middle of the fifteenth century, China, a nation then of about 100 million people—which, as Robert Temple has pointed out, had already invented gunpowder, the compass, paper money, the stirrup, the decimal system, and the seismograph—had walled itself off physically, psychologically, and economically from the Western world. With medicine, as in other areas, the Europeans would have to work things out for themselves. In his chapter on Chinese medicine, Roy Porter wrote:

> From the wider perspective there is a key difference between the eastern and western medical traditions. Both initially shared common assumptions about the balanced and natural operations of the healthy body and these were inscribed in hallowed texts. Western medicine alone radically broke with this. An entirely new medicine grew up in the West—scientific medicine—building upon the new sorts of knowledge, programme and power that followed from anatomy and the investigations of the body it opened.

Contrasting early Greek and early Chinese science, Geoffrey Lloyd and Nathan Sivin (scholars of each, respectively) together wrote *The Way and the Word* (2002), a book in which they discuss how knowledge of the natural world was acquired, propagated, and disseminated in each of the two cultures. Greek scholars "focused on nature and on elements, concepts that seem familiar and obvious to those educated in modern science," but the Chinese had an altogether "different set of fundamental concepts, not nature and the elements, but the *tao*, *ch'i*, yin-yang, and the five phases. Where Greek inquirers strove to make a reputation for themselves as new-style Masters of Truth, most Chinese Possessors of the Way had a very different program, namely to advise and guide rulers." Because the realms of heaven, Earth, and man were united in the person of the emperor, the role of his advisors was to keep these elements in balance; a virtuous emperor meant that no untoward calamities would befall heaven or earth. Treatises like the *Huang-ti Nei Ching*, therefore, were as much cosmological as medical, while comparable Greek scholars, such as Hippocrates (who did not seek a position at court but rather strove to have his ideas accepted over those of competing teachers), with no patrons to please, had to "fall back on their own resourcefulness in building a reputation and in making a living."

Among the concepts that didn't make it successfully into China until much more recently was that of medical bacteriology—the idea that disease is caused by miniscule invasive creatures. Somewhere around 50 BC, the Roman writer Varro (116–27 BC) associated fever with "marsh insects," though probably not with flies and mosquitoes: "in swampy places minute creatures live that cannot be discerned with the eye and they enter the body through the mouth and nostrils and cause serious diseases," he wrote. Although most Europeans believed that bubonic plague, which ravaged Europe from 1347 to 1350 and arose sporadically until 1670, was caused by bad air (miasmas) or bad faith, there was an inkling of belief that it might be spread by contagion. (The word malaria also means "bad air.") While besieging the city of Caffa, on the Black Sea, in 1347, Tartars catapulted corpses of plague victims over the walls of the city—the first recorded instance of biological warfare. According to Paul Ewald's 1994 *Evolution of Infectious Disease*, in 1546, Girolomo Fracastro published the idea that diseases were caused by disease-specific germs that could multiply within the body and be transmitted directly from person to person or indirectly on contaminated objects; moreover, he proposed that variations in the intensity of epidemics could be attributed to changes in the virulence of germs. It would be another three hundred years, however, before the germ theory of disease would become prevalent in the West.

What caused diseases in the early Chinese view? In the recent textbook *Basic Theory of Traditional Chinese Medicine* (Wu 2002), the chapter "Causes of Disease" begins with this definition:

> The pathogenic factors in TCM can be divided into four categories: (1) exogenous pathogenic factors including six climatic factors and pestilence. (2) endogenous pathogenic factors including seven emotions, improper diet and overstrain, etc. (3) secondary pathogenic factors including phlegm, retention of fluid and blood stasis. (4) other pathogenic factors including various traumatic injuries, injuries due to physical and chemical factors and injuries caused by insects and animals.

In other words, disease might be caused by an emotional breakdown, improper diet, overwork, retention of fluids, or any number of factors that can, in fact, precipitate illnesses, but because the germ theory of disease had not yet been developed, the Chinese—as well as Westerners—attributed disease to everything but pathogens, the actual cause of many diseases. For example, in the 1741 *Chou hou pei chi fang* cited in Joseph Needham's 2000 study of medicine in China we read that:

> Smallpox is a congenital poison rooted in the conjugation of yin and yang at the very beginning of conception. Once the temporal cycles of susceptibility to illness and the seasonal epidemic *chhi* stimulate it, the disease will unfailingly break out. If one waits to deal with it until it has broken out, because epidemic *chhi* is already rampant the symptoms will generally be unmanageable.

Although without the germ theory, the Chinese nevertheless seem to have invented *variolation*, the practice of implanting live variola into an incision, which often resulted in a milder form of the disease with a much lower fatality rate than if the disease had been transmitted through the respiratory tract. Robert Temple tells us in *The Genius of China*, "The technique first came to public attention when the eldest son of the Prime Minister Wang Tan (957–1017) died of smallpox. Wang desperately wished to prevent its happening to other members of his family, so he summoned physicians, wise men and magicians from all over the Empire to try to find some remedy. One Taoist hermit came from O-Mei Sand, and brought the technique of inoculation and introduced it to the capital." Because it was such a devastating disease, there were more than fifty ancient Chinese treatises written on smallpox, and according to Joshua Horn (1969), "by the sixteenth century, over 200 years before Jenner's epoch-making discovery, a form of inoculation against smallpox, consisting of extracting and drying the contents of a pustule from a smallpox victim and blowing the powder into the nose, gained wide acceptance in China." The Chinese recognized that survivors of smallpox became immune, and physicians tried to infect healthy people in the hope that a mild infection would create immunity.

The Western version of variolation came later, probably not influenced by the Chinese. In 1715, English noblewoman Lady Mary Wortley Montagu had survived an attack of "the speckled monster," but she was permanently scarred and her eyelashes had fallen out. Her husband had been appointed ambassador to Constantinople, and while there, she had her own children successfully inoculated. In

1721, Lady Montagu introduced the idea of "inoculation" to England, and in what became known as the "Royal Experiment," six condemned prisoners at Newgate Prison were inoculated and promised a full pardon if and when they recovered. They all became free men. With variolation, the fatality rate was reduced from 30 percent to about 1 percent. And the procedure spread to America, where John Adams was successfully variolated in 1764 and Thomas Jefferson likewise in 1766. It was not until 1796, when Edward Jenner inoculated eight-year-old James Phipps with cowpox (which was harmless to humans) and found that the lad became immune to smallpox that vaccination was accepted in England and western Europe. Western belief in person-to-person transmission (as distinct from miasmatic transmission) was also strengthened by the introduction of smallpox to Europe by returning crusaders and by the later introduction of syphilis to Europe by travelers to the New World.

The interesting parallels of variolation aside, while Chinese medicine maintained continuity over the centuries to the present, the further development of Western medicine of bacteriology led to increasing divergence between the two practices. In Europe, the Dutchman Antoni van Leeuwenhoek (1632–1723) was the first to discover bacteria, free-living and parasitic microscopic protists, sperm cells, blood cells, microscopic nematodes and rotifers, and much more. Leeuwenhoek had learned to grind lenses, made simple microscopes (though the more powerful compound microscopes had already been invented), and began observing with them. On September 17, 1683, he wrote to the Royal Society of London about his observations on the scurf between his own teeth, "a little white matter, which is as thick as if 'twere batter." He repeated these observations on two ladies (probably his own wife and daughter) and on two old men who had never cleaned their teeth in their lives. Looking at these samples with his microscope, Leeuwenhoek found "an unbelievably great company of living animalcules, a-swimming more nimbly than any I had ever seen up to this time. The biggest sort . . . bent their body into curves in going forwards. . . . Moreover, the other animalcules were in such enormous numbers, that all the water . . . seemed to be alive." These were among the first observations on living bacteria ever recorded.

Experimenting with bacteria in the mid-nineteenth century, the French chemist Louis Pasteur (1822–95) conclusively disproved the theory of spontaneous generation (which held that living organisms could arise from nonliving substances) and led the way to the germ theory of infection. But even by the end of the nineteenth century, when European medical advances had become well known (and well publicized) in the West, traditional Chinese practitioners had not acknowledged germ theory and were still attributing disease simply to an imbalance of the body's humors.

German bacteriologist Robert Koch (1843–1910) made a number of advances in understanding of germs and codified the principles of germ study. He identified the bacterial causation of anthrax disease and demonstrated publicly the life cycle of the anthrax bacillus, and he described the significance of spores in transmission of the disease; his study, published in 1877, well before the century was out, was a landmark in the history of the germ theory. Five years later, Koch revealed before the Berlin Physiological Society the bacillus that causes tuberculosis, *Mycobacterium tuberculosis*, and in 1883, in India, he demonstrated that *Vibrio cholerae*, the bacillus that causes cholera, was communicated by polluted water. Koch's four postulates form the basis of modern bacterial epidemiology:

1. The specific organism should be shown to be present in all cases of animals suffering from a specific disease but should not be found in healthy animals.
2. The specific microorganism should be isolated from the diseased animal and grown in pure culture on artificial laboratory media, over several generations.
3. The freshly isolated microorganism, when inoculated into a healthy laboratory animal, should cause the same disease seen in the original animal.
4. The microorganism can be retrieved from the inoculated animal and cultured anew.

By the early twentieth century, many of the common killer diseases—smallpox, plague, influenza, whooping cough, measles, tuberculosis—

were known to be caused by pathogens, and yellow fever and malaria were known to be caused by microbes delivered by mosquitoes. It would have been impossible for Chinese doctors to ignore this information, and only die-hard traditionalists could continue to prescribe herbs or acupuncture for killers such as influenza, measles, and tuberculosis. (But, as we shall see, a cure for malaria came straight out of the pharmacopoeia of China.)

Herbal and animal-based substances continue to be prescribed in China for a wide range of conditions, based largely on the ancient principles of TCM. Among the more commonly treated disorders are skin diseases, including eczema, psoriasis, acne, and rosacea; gastrointestinal disorders, including irritable bowel syndrome, chronic constipation, and ulcerative colitis; gynecological conditions, including premenstrual syndrome, dysmenorrhea, and infertility; respiratory conditions, including asthma, bronchitis, chronic coughs, and allergic and perennial rhinitis and sinusitis; rheumatological conditions such as rheumatoid arthritis; urinary conditions such as chronic cystitis; psychological problems such as depression and anxiety; and chronic fatigue syndrome. Chinese herbals describe the use of every plant from ginseng and alfalfa to sassafras, cloves, myrrh, frankincense, cannabis, parsley, sage, rosemary, and thyme; and of course, various parts of various animals are still high on the list of curative substances. A list of the substances used in TCM and the conditions for which they are applicable occupies most of the 532-page *Fundamentals of Chinese Medicine*.

There was one widespread and deadly disease for which TCM provided no treatment; indeed, the Chinese government was even reluctant to acknowledge its existence. It was not until August 2001 that the Chinese government finally admitted that the country was facing a crisis with AIDS, a disease that has already claimed 23 million lives around the world and appears to be on the rise in heavily populated countries such as China and India. Chinese government officials now estimate that the total number of people infected with HIV is about six hundred thousand, but many experts working in China believe that even this figure is a serious underestimate and that the epidemic is much more widely spread, with the true number of people living with HIV being closer to 1.5 million. There is no known cure for HIV or AIDS, either in Western or traditional Chinese medicine, but researchers have found that TCM can be of particular use to persons with HIV who are experiencing symptoms of diarrhea, loose stools, weight loss, abdominal pain, nausea, headaches, and enlarged lymph nodes. Studies in China and Thailand, where TCM has almost universal acceptance and a record of use going back several thousand years, have found that many people have successfully used TCM for addressing other HIV factors including fatigue, general energy loss, and declining mental powers.

In addition to some relief from HIV symptoms, some promising results have been obtained in TCM treatment of Hepatitis C. What does TCM say about cancer? One TCM Web page that I consulted noted that cancer can be caused by air pollution, food, and radiation—in agreement with Western medicine. The Cameron Clinic of Chinese Medicine (www.camclinic.com) says that cancer, like other diseases. is a manifestation of imbalance or disharmony in the body:

> In Chinese Medicine, illness is an energy imbalance, an excess or deficiency of the vital substances. Qi or vital energy controls the body's working as it travels along the channels or meridians. A person is healthy when there is a balanced, sufficient flow of Qi, which keeps the blood and body fluids circulating and fighting disease. If Qi is blocked for any reason or is excessive or deficient, disease may result. The flow of energy may be disrupted by an unblanced diet or lifestyle, overwork, stress, repressed or excessive emotions, lack of exercise, external pathogenic factors, etc. . . . The Chinese medicine practitioner determines the individual's pattern of disharmony rather than a condition such as breast cancer or colon cancer. The prescribed treatment depends on the individual's specific imbalance. Once the pattern of disharmony is identified a treatment plan is formulated to restore balance.

On the "Traditional Chinese Medicine Information Page" (www.tcmpage.com), I found this discussion of breast cancer:

> Chinese medicine believes that the fundamental cause of breast cancer is emotional disturbances such as excessive thinking or anger, which lead to functional disorders of the Liver and Spleen. A com-

mon causative pattern is that excessive Heat from a deficient Liver, combined with Phlegm Dampness due to Spleen dysfunction, results in the blockage of Chi and Blood, which then "condenses" into breast cancer. Another common causative pattern is when Liver Deficiency and Kidney Deficiency lead to Chi and Blood Deficiency. Chronic Chi and Blood deficiency then leads to Qi Stagnation and Blood Stasis, which causes the formation of lumps in the breast. A third pattern is when Qi Stagnation and Phlegm accumulation lead to excessive Heat toxins, which then turn to hard breast lump masses.

Cancer has always been one of humankind's most frightening diseases. It has been found in fossilized bone tumors in ancient Egyptian mummies, and the term *cancer*, which means "crab" in Greek, was coined for the disease by Hippocrates because of the spreading, fingerlike projections of some visible cancers. Hippocrates believed that an excess of black bile in various organs caused cancer, and Galen's misguided theories of the causation of disease held sway until the revised teachings of Paracelsus overturned them in the sixteenth century. Despite theories that cancer was caused by irritation, trauma, and parasites (a Nobel Prize was awarded in 1926 to Johannes Fibiger of Denmark who demonstrated that cancer in mice was caused by a worm), clinical experimentation has shown that the disease could be caused by carcinogens such as coal tar, benzene, aniline dyes, asbestos, radiation (including sunlight), hydrocarbons (particularly in tobacco smoke), and, most recently, viruses.

In any discussions, Chinese or Western, the reasons that cancer strikes some people and not others is largely unknown. We know that cancer develops when cells in a part of the body begin to grow out of control, and instead of dying, as normal cells do, cancer cells continue to form abnormal cells. The process known as *metastasis* occurs when cancer cells get into the bloodstream or lymph nodes and begin to replace normal tissue. Cancer cells develop because of damage to the cell's DNA, but exactly how this happens is not clear. Galen considered cancer incurable, but by the early twentieth century, the (sometimes successful) removal of tumors by surgery was the prevailing treatment in Western medicine. Hormone therapy and radiation followed, and chemotherapy was discovered to kill proliferating cancer cells by damaging their DNA. It has now been shown that a diet high in vegetables, fruits, whole grains, and beans, when combined with regular exercise, reduces the risk of cancer. Not smoking, staying out of the sun, and avoiding excessive radioactivity (such as nuclear explosions) do the same. If a balanced diet can reduce the risk of cancer, it now appears that the TCM theory about correcting an imbalance of yang and yin can presumably also have validity after all. Arguments about risk and cause aside, the application of the right animal, vegetable, or mineral pharmaceuticals could probably cure a number of diseases or ameliorate the symptoms, even if the explanation is 2,000 years old.

For example, *Artemesia annua*, known in English as sweet wormwood and in Chinese as *qinghao*, has been shown to be a cure for malaria. As Andy Coghlan wrote in a 2003 article in *New Scientist*, "a hitherto unknown but vital weakness in the malaria parasite has been exposed by studying extracts from ancient Chinese anti-fever remedies." This plant contains the active ingredient artemisinin, which kills the parasitic protozoan *Plasmodium falciparum*, transmitted to humans via the malaria mosquito, *Anopheles gambiae*. It was only in a recent study by scientists at St. George's Hospital Medical School in London (Eckstein-Ludwig et al. 2003) that the way in which artemisinin kills *Plasmodium falciparum* was identified, but the molecular composition of *Artemisia* was not important to early Chinese practitioners—they just knew it worked.

In an article in *Science* in 1985, Daniel Klayman introduced "*Qinghaosu* (Artemisinin): a new antimalarial drug from China." The drug may have been new to Western medicine, but *qinghao* (sweet wormwood) was known in China as far back as 168 BC, when, wrote Klayman, "its earliest mention occurs in the *Recipes for 52 Kinds of Diseases*, found in the Mawangdui Han dynasty tomb dating from 168 BC." The plant is also mentioned in the AD 340 *Zhou Hou Bei Ji Feng* (Handbook of Prescriptions for Emergency Treatments), which provides a recipe for *qinghaosu* (pronounced "ching-how-sue") in an infusion for treating fever. More than 1,200 years later, Li Shih-chen realized that *qinghaosu* could be used for treating the symptoms of malaria and included the treatment in the 1597 *Pen Ts'ao Kang Mu.* There things sat until 1972, when Chinese scientists successfully extracted the plant's active compound, calling it *qinghaosu*—which was *artemisinin* in conventional scientific terminology, after the plant's Latin name, *Artemesia annua.*

Since the discovery of artemisinin, wrote Klayman, "the compound has been used successfully in several thousand malaria patients in China, including those with both chloroquine-sensitive and chloroquine-resistant strains of *Plasmodium falciparum.*" (Chloroquine is a traditional treatment for malaria.) Further studies in China and Vietnam have confirmed that it is a highly effective compound with close to 100 percent response rate for treating malaria. It has the ability to destroy the malaria parasite by releasing high doses of free radicals that attack the cell membrane of the parasite in the presence of high iron concentration. In fact, over one million malaria patients have been cured via this method. Their symptoms also subsided in a matter of days.

Research is now being conducted on claims that *Artemisia* may also kill cancer cells. So far, the most extensive study on the use of artemisinin as an anticancer agent has been carried out by bioengineering scientists Narenda Singh and Henry Lai of the University of Washington, as reported in 2001 in the journal *Life Sciences.* They concluded that artemisinin kills malaria but it can also be used to treat various cancers. Singh and Lai wrote, "Since it is relatively easy to increase the iron content inside cancer cells in vivo, administration of artemisinin-like drugs and intracellular iron-enhancing compounds may be a simple, effective, and economical treatment for cancer."

Iron is required for cell division, and it is well known that many cancer cell types selectively accumulate iron for this purpose. Most cancers have more iron-attracting transferring receptors on their cell surface than normal cells. In laboratory studies of radiation, resistant breast cancer cells that have a high propensity for accumulating iron revealed that artemisinin shows 75 percent cancer-cell-killing properties in eight hours and almost 100 percent killing properties within twenty-four hours when these cancer cells are "preloaded" with iron after incubation. On the other hand, the normal cells remained virtually unharmed. In *Fundamentals of Traditional Chinese Medicine,* sweet wormwood (*Artemisia*) "clears heat, resolves summerheat . . .

treats warm disease, fever, malaria, dysentery, jaundice, scab, and itching."

Klayman's article on *qinghaosu* concludes with the following words: "Scientists in the People's Republic of China have not only contributed to a structurally novel and well-tolerated class of rapidly acting anti-malarial agents but have also encouraged the investigation of folk medicine." But because artemisinin derivatives are expensive to produce and degrade quickly, a synthetic form was developed by University of Nebraska pharmacologist Jonathan Vennestrom and his colleagues that was cheaper to make and easily manufactured.* Like natural artemisinin, the synthetic form kills malarial parasites by producing free radicals, and although it is still being tested, it's possible that "a three-day dose should be sufficient to completely cure malaria" (Avasthi 2004). Perhaps, in this instance at least, the combination of TCM and modern science can accomplish what neither could accomplish separately.

There is, then, some overlap in the principles and practices of Chinese and Western medicine. While an arthritis patient in New York who saw a Western-trained doctor would probably not find himself with a prescription for alfalfa, nutmeg, or horseradish, more and more Westerners are recognizing the known beneficial properties of some plant constituents and have learned of positive uses of acupuncture, thereby crossing over to at least some of the precepts of traditional Chinese medicine. Although acupuncture had a small following in nineteenth century France and Britain, only in the last generation has it made great inroads into some aspects of Western treatment. This is due partly to a new multiculturalism and partly to rejection in some quarters of high-tech values; but it also results from explanations of acupuncture anesthesia in terms of endorphins and other neurotransmitters.

Many Westerners choose the herbal route because of its reliance on "natural" materials as opposed to those manufactured in a laboratory.

* In *Science* (January 7, 2005), however, Martin Enserink reports that there aren't nearly enough of the new drugs to go around, and artemisinin-based therapy is too expensive for the malaria patients, particularly in Africa, who need it most. In December 2004, the Bill and Melinda Gates Foundation announced a $40 million investment to develop a bacterium to churn out artemisinin, but it unfortunately may take years for this to be accomplished.

Herbs are typically prescribed in combination, which herbal practitioners believe balances them, allowing them to act together in a way that increases their efficacy. As with all traditional Chinese medications, herbal medicine seeks primarily to correct internal imbalances rather than to treat symptoms alone, and therapeutic intervention is designed to encourage the self-healing process. Of course, Western medicine is not the polar opposite to this. Using medication to correct imbalances—in thyroid production or vitamin deficiencies—is standard practice. And not all Western medications are synthesized "unnaturally" in the lab; quinine, a long-standing antimalarial medication, is derived from the bark of the cinchona tree (*Cinchona pubescens*), while scopolamine, used extensively for motion sickness, comes from the jimsonweed, *Datura stramonium*. Furthermore, some plants that are known to have beneficial medicinal effects are employed by both schools: the clear gel from the leaves of aloe vera is used for healing wounds and burns; cloves are used for flavoring food, but they are also used in Western medicine and TCM as anesthetics and antispasmodics; eucalyptus leaves are used around the world to relieve coughs and colds; belladonna (also known as "deadly nightshade" because it is poisonous) contains atropine, used in conventional medicine to dilate the pupils for eye examinations and as an anesthetic and in TCM to relieve intestinal colic, treat peptic ulcers, and relax distended organs, especially the stomach and intestine.

Despite such overlaps in the principles and practices of Chinese and Western medicine, many differences remain, largely because of the rationales that created them and in the way they developed. For some centuries, practitioners of early European and Chinese medicine were nonetheless roughly analogous, emphasizing herbal treatments and a number of humors or elements that had to be kept in equilibrium. Joseph Needham (2000) identifies the strengths and weaknesses of the two systems:

> When we turn to look at traditional Chinese medicine, we have to recognize at once that the concepts with which it works—the yin and the yang, and the Five Elements—are all more suited to the times of Hippocrates, Aristotle and Galen, than to modern times.

> . . . A feature in which traditional Chinese medicine is extremely good is its organic approach to illness. Two patients with identical symptoms may be given quite different treatments, depending on their backgrounds, which the physician has enquired about, and the general pictures of their body processes as ascertained in the examination. Another excellent feature of traditional Chinese medicine is the notion of disease as a process that passes through several stages. This can lead to some very sophisticated cures. Generally speaking, a strength of traditional Chinese medicine lies in curing chronic diseases.

How innovations were absorbed into the two systems made a substantial difference in the way medicine was practiced. The organization of the profession in Europe around medical schools may have been decisive in producing more systematic responses to new diseases. Hospitals provided opportunities for repeated observations of the patient's symptoms, whereas Chinese physicians often visited the patient once and provided a course of treatment. In a hospital, a cure that worked once could be tried again on the next patient, and professional colleagues were on hand to observe the result. European practitioners, therefore, were encouraged to try new cures. If the cures didn't work, they might then try something else. It is not, therefore, surprising that European doctors reacted to new diseases by altering major elements of the older theory and practice. By contrast, Asian medical experts, who did not operate in hospital environments, met new disease experiences by holding fast to ancient authorities—or claiming to do so even when something new crept in (McNeill 1996).

Of the substances prescribed potions work as well, say, as aspirin, but some probably don't work at all, relying though, in some cases, on the "placebo effect," where symptoms are alleviated because the patient believes he is receiving medication that will alleviate his symptoms. Aspirin, which is composed of acetylsalicylic acid, is an important element in modern Western medicine, but for a long time its employment was closer to that typical of a TCM substance. People took aspirin simply because they knew it worked. Only recently have scientists begun to understand its workings and its range of medical effects, ranging from pain relief to reducing the likelihood of blood clot formation.

There is no question that prescriptions in Western medicine often do not accomplish what they are prescribed for, or that even with a substance as ubiquitous as aspirin, there are sometimes unexpectedly serious side effects. In contrast, there is a popular impression that TCM therapies, even if they don't always work, at least do no harm. After all, they are derived from "natural" substances, and how much harm can herbs, plants, or powdered animal parts do? Lots, as it turns out. Eating the bulbs of hyacinth, narcissus, or daffodils can be fatal; the branches of oleander are extremely poisonous; the leaves of foxglove can cause dangerously irregular heartbeats; all parts of laurels, azaleas, and rhododendrons are deadly; jasmine, mistletoe, and yew berries can kill you; and everybody knows what Socrates drank to commit suicide.

There are many people in Asia and elsewhere who believe that "alternative" medical treatments and practices are superior to those of Western medicine, which typically rely on artificial drugs and focus on repairing damage only after it has occurred. So it may come as a surprise to learn how enormously popular "unconventional" medicine is in the United States. A survey conducted in 1993 by David Eisenberg and his colleagues found that 33.8 percent of the respondents reported using at least one unconventional therapy, including herbal medicines, massage, megavitamins, folk remedies, and homeopathy, in the past year. In a follow-up survey, published five years later, Eisenberg and colleagues found that the percentage of those using an alternative therapy had increased to 42.1 percent, while the number of users visiting an alternative medicine practitioner jumped from 36.3 percent to 46.3 percent.

Doctors of Western medicine must be government certified to practice, but, with the exception of osteopathy or chiropractic, holistic practitioners are largely unregulated and do not have to be certified or even trained. Almost anybody can set himself up as a "therapist," and many "natural" health products are unregulated and marketed simply as food supplements. The supplements may thus be mis- (or self-) prescribed, and some may contain undetected toxic material. Debora MacKenzie in 1998, for example, told of 2,000 women in Belgium who signed up for slimming treatments, but because the wrong herb was prescribed, more than 120 of them developed kidney failure. A study conducted by Robert Saper and his colleagues, including Dr. Eisenberg, in the *Journal of the American Medical Association* in 2004, found that "one of five Ayurvedic herbal medical products, produced in South Asia, and available in Boston area stores, contains potentially harmful levels of lead, mercury, and/or arsenic." (Ayurvedic medicine originated in India more than 2,000 years ago and relies heavily on herbal medicine products.) Eisenberg, a professor at Harvard Medical School, commented, "In order to investigate the efficacy of commonly used dietary supplements including Ayurvedic remedies we need to test high-quality standardized products free of contaminates and dangerous toxins. . . . Over-the-counter herbs and supplements with high levels o heavy metals are simply dangerous."

Traditional Chinese medicine has had a long and noble history over two millennia of Asian history, and in many parts of the world it is actively practiced today. There is no question that many components of the TCM pharmacopoeia are successful in suppressing fever, reducing swelling, curing headaches, nausea, dizziness, and toothache, eliminating pain, assuaging the agony of gallstones, or easing childbirth. TCM probably cannot cure cancer, heart disease, AIDS, tuberculosis, cholera, typhoid, typhus, dengue fever, influenza, measles, or chickenpox. There does not appear to be any evidence to support the claims that various substances listed in the Chinese materia medica can enhance one's virility, detect poisons, or make one live longer.

There are arguments to be made—not all of them convincing—that it is necessary (and morally acceptable) to use animals like mice, rats, rabbits, or monkeys in tests that might have beneficial applications for human medical needs. These animals are sacrificed for a "higher purpose," namely the production and testing of vaccines, hormone preparations, or even cosmetics. The number of laboratory mice and rats that die every year in the name of medical research must be astronomical. Within the precepts of today's TCM, however, some animals are killed to provide what practitioners are convinced are cures for ailing people, or people who might be sexually dysfunctional. In many cases, these prescriptions do not work, or do not work as well as some synthetic pharmaceuticals.

But it is not, after all, the use of animal parts *per se* that is the problem—it is the slaughter of the animals for what might be specious applications, or worse, the slaughter of critically endangered species.

Not all animal-based prescriptions of TCM require that an animal die. There are some parts of some animals that may or may not work as pharmaceuticals, but at least do not require that the animal be killed. "Antler velvet," the soft covering of maturing deer antlers, for example, appears to be the TCM analog of aspirin. Deer of all species begin their annual antler growth in the spring when their antlers are soft and covered in a thin skin, which bears short, fine hairs and resembles velvet. The growing antlers are warm to the touch and very sensitive. By late summer the antlers have attained their maximum size, and the thin skin

Package of deer antler. The large tuber to the left of the tree is ginseng. This packet was bought in a shop in New York City's Chinatown that also sold dried sea horses and a variety of herbal preparations. Deer antler and deer antler velvet are said to cure joint stiffness and arthritis, boost energy levels, aid muscle recovery, balance cardiovascular activity, strengthen the immune system, increase libido, and heighten general vitality.

with the velvety hair dries, loosens, and is rubbed off on shrubs and small trees, leaving the bony antlers uncovered. Because the "velvet" is supplied with blood vessels and contributes to the growth of the antlers, it is not surprising that for centuries, Chinese practitioners attributed special qualities to this substance. For example, the *Pen Ts'ao Kang Mu*, written by Li Shih-chen in 1597, gives this as the medicinal properties of antler velvet, known as *Lu rong*:

> For vaginal bleeding, convulsions with feverish cold. It benefits the vitality and strengthens the mind. It assists the growth of permanent teeth. A good tonic for weak people. For arthritis and backache. It is a diuretic. . . . For vesicular calculi, osteomyelitis. To quieten the placenta. For nymphomania, menorrhagia.

Was this just some weird sixteenth-century prescription? Hardly. Here's what the 1986 *Chinese Herbal Medicine Materia Medica* says about the same substance:

> Tonifies the governing vessel, augments essence and blood, and strengthens sinews and bones; used especially in cases of deficient essence and blood in children with such physical and/or mental developmental disorders as failure to thrive. Mental retardation, learning disabilities, insufficient growth, or skeletal deformities (including rickets).

A commercial Web site advertising antler velvet says, "velvet antler is a natural nutritional supplement used to relieve joint stiffness, increase bone density, boost energy levels, aid muscle recovery, balance cardiovascular activity, strengthen the immune system, increase libido activity, and heighten general vitality. Today, new research has highlighted the supplement's powerful benefits for osteoarthritis, rheumatoid arthritis, and osteoporosis sufferers."

In order to obtain the velvet for medicinal uses (it is prepared as slices through the antler, which look like poker chips, or dried and ground into a powder) the deer does not have to be killed. The sika deer (*Cervus nippon*), common in Japan and China, is widely farmed in China and various parts of the former Soviet Union for antler velvet production, wrote Valerius Geist in 1998. The deer are raised on special farms where, in the spring, their velvet antlers are sawed off; and as with all deer, the antlers will grow back the next year.

Plants can be picked; leaves can be plucked, fruits, vegetables, and nuts can be harvested, and seeds can be collected without threatening the species, but some animals must be killed in order that the valuable parts can be harvested for use in TCM. To obtain shark fins, the sharks must be killed, and because sea horses are used in their entirety, they are killed wholesale. You cannot remove the bones or skin of a tiger, or the gall bladder of a bear, without killing the animal, and while some clever entrepreneurs have figured out how to obtain bear gall from a living bear, the technique is so awful that death for the bear might be preferable. It is possible to anesthetize a rhino and saw its horn off, but it is a difficult, cumbersome, and dangerous process that might result in the death of the rhino anyway, so poachers take the easier path and just shoot the rhino. So that some might make use of their fins, bones, horns, or gall bladders, some of our most charismatic wild animals are in danger of being wiped off the face of the earth.

4

Horn of Plenty

As we've seen, there are many parallels between the development of Western and Chinese medical practices. Both systems were holistic in that they involved the treatment of the whole person, not just the symptoms, and sought somehow to restore the natural balance of the body. Similar systems of treatment developed, where maladies were often treated with specific preparations derived from plants and animals that might be found in the garden, the barnyard, or the surrounding countryside. But because some exotic conditions called for exotic remedies, medical practitioners sought unusual animals whose parts might have uncommon curative powers. There are few animals stranger than a rhinoceros, a lumbering giant with horns growing where no other animal has horns, so it is not surprising that rhino horn became an integral part of early Chinese medicine. The use of rhino horn, however, can be traced to the unicorn, another animal with a horn growing from a totally unsuspected place.

The Fabulous Unicorn

There never actually *was* a unicorn, but many people thought there was, and that's almost as good. "There is no doubt," wrote Willy Ley in 1948, "that the unicorn is the most glorious of all the mythical creatures to be found in books before and also after Pliny." Before Pliny, a Greek physician named Ctesias, in the service of Darius II, King of Persia, returned to his homeland in 398 BC and wrote *Indica*, a book containing stories he had collected during his seventeen years in Persia. One of his tales describes certain wild asses that are as large as horses, and larger. Their

This spirited beast appeared in Conrad Gesner's 1551 *Historia Animalium* in the days when people had no difficulty believing in the existence of unicorns.

bodies are white, their heads dark red, and their eyes dark blue. They have a horn on the forehead about a foot and a half in length. The dust filed from this horn is administered in a potion as a protection against deadly drugs. The base of the horn, for some two hands' breadth above the brow, is pure white; the upper part is sharp and of a vivid crimson; and the remainder, or middle portion, is black. Those who drink out of these horns, made into drinking vessels, are not subject to the holy disease (epilepsy). Indeed, they are immune even to poisons if, either before or after swallowing such, they drink wine, water, or any liquid from these beakers.

After Ctesias—and probably based largely upon his description—other authors included the unicorn in their studies of natural history. Aristotle, about a half—century later, asserted, "There are . . . some animals that have one horn only, for example, the oryx, whose hoof is cloven, and the Indian ass, whose hoof is solid. These creatures have a horn in the middle of their head." Then there was Pliny (the Elder), who was born in AD 23 and killed in 79 at Pompeii while trying to observe the eruption of Mt. Vesuvius. In *Historia Naturalis*, considered one of the most important of all early natural history books, Pliny makes no mention of any special powers of the horn but tells us, "The Orsaean Indians hunt an exceedingly wild beast called the monoceros, which has a stag's head, elephant's feet, and a boar's tail, the rest of the body being like that of a horse. It makes a deep lowing noise, and one black horn two cubits [about 40 inches] long projects from the middle of its forehead. This animal, they say, cannot be taken alive."

Somehow, a vaguely horse-shaped quadruped with a single horn managed to insinuate itself into the world's natural (and unnatural) histories, without what we might today consider a necessity—a direct sighting. Of course, much of ancient natural history was based on twice-told tales, because so few people were able to travel far in search of wild animals they had heard about. As Odell Shepard put it in his brilliant study of the unicorn, "The fact that no one ever saw a unicorn did not disturb belief in the slightest degree. No one in mediaeval Europe ever saw a lion or an elephant or a panther, yet these beasts were accepted without question upon evidence in no way better or worse than that which vouched for the unicorn." A unicorn wasn't really such a strange idea anyway; it was just an animal with a single horn where most other hoofed animals have two. There are a lot of mammals much more unlikely: the platypus, the giraffe, the elephant, the kangaroo, the armadillo, the anteater, the manatee and the narwhal are much too weird to exist—and yet they do. Indeed, there is a living animal that does have a single horn—or maybe even two—growing out of its nose, and few people question the validity of the rhinoceros.

There are several places in the Bible where unicorns are mentioned: Numbers 23:22 says, "God brought them out of Egypt; he hath as it were, the strength of an unicorn." And Job 39:9–11: "Will the unicorn be willing to serve thee, or abide by thy crib? Canst thou bind the unicorn with his band in the furrow? Or will he harrow the valleys after thee? Wilt thou trust him because his strength is great?" "One thing is evident in these passages," writes Shepard. "They refer to some actual animal of which the several writers had vivid if not clear impressions.

. . . Nothing about it suggests that it was supernatural, a creature of fancy, for it is linked with the lion, the bullock and the calf, yet it was mysterious enough to inspire a sense of awe, and powerful enough to provide a vigorous metaphor." In later years, the absence of the unicorn would be attributed to its missing the ark and drowning in the flood.

Long before the unicorn achieved its preeminence in Europe, it was thriving in China. In his 1977 *Unicorn: Myth and Reality*, Rüdiger Robert Beer asked if China could have been its original home: "The first notice of the beast in China has been placed as far back as 2697 BC. Extant descriptions of *chi-lin*, the Chinese unicorn, remind one of European conceptions of the animal: a body like an axis deer, horse's hoofs, an oxtail, as well as a horned, wolf-like head. King of the 360 animal species then recognized, *chi-lin* was reputed to reach 1000 years of age." An exhibition of treasures from the Silk Road, mounted by New York's Asia Society in October 2001, featured no fewer than three unicorn sculptures from the pre-Han and Han dynasties, beginning in 206 BC and ending in AD 221.

On her marvelous Web site, "The Mythic Chinese Unicorn *Zhi*," Jeannie Thomas Parker of the Royal Ontario Museum introduces the viewer to a wooden carving depicting a powerful animal, built along the lines of a fighting bull, with a single, tapered horn protruding from its forehead; it was also found in Gansu Province, dated from the Han Period, about 200 BC. This figure does not represent *chi-lin* at all, she writes, but rather *zhi*, a one-horned, goatlike creature that had long been represented in pictographs and statues of a male and female guarding the entrances of the courts of law; later, they were used as guardians of underground burial chambers. "Implacable and incorruptible, the tomb guardians *zhi* were intended to serve through all eternity to avert or ward off any bad influences or evil spirits that might attempt to violate the underground abode of the deceased."

In Pleistocene times, twenty thousand years ago, gigantic, one-horned rhinoceroses roamed throughout Europe and Asia, but with a more traditional placement of the horns—on the nose. The woolly rhinoceros, *Coelodonta antiquitatis*, for example, inhabited the steppes of Europe. It stood nearly six feet high at the shoulder and resembled the modern white rhino in general proportions, but if the drawings on the walls of Chauvet Cave in France are to be believed, these rhinos of the Ardèche had enormous horns. Discovered in 1994, the walls of Chauvet are covered with more than three hundred paintings of horses, lions, bears, mammoths, hyenas—and spectacular long-horned rhinos. In their 1996 book about the caves, Jean-Marie Chauvet (for whom the cave was named), Étienne Brunel Deschamps, and Christian Hillaire wrote:

> Chauvet cave was startling in many respects—no decorated cave of such size had ever been found in this part of France; it was completely intact, with its varied traces of human and animal visitors . . . and it has so many images of so many different species, most notably rhinoceroses, big cats, and bears—animals that were hitherto unknown in this region's Ice Age art, but were also rare anywhere else, and certainly not depicted with such prominence on main panels. For example, less than twenty rhinos were previously

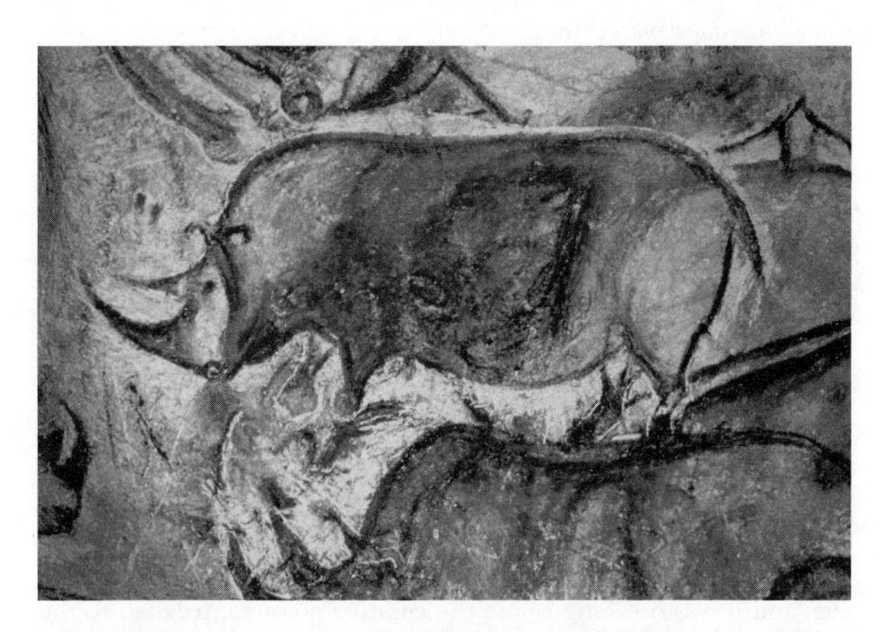

As long as thirty thousand years ago, rhinos walked the plains of Europe. This drawing was found on the wall of Chauvet Cave, in the Ardèche region of southeastern France.

known in European cave art, whereas Chauvet contains two or three times that many.

The art of Chauvet has been dated at approximately thirty thousand years of age; the same approximate date applies to a hardy band of Stone Age people that settled in northern Siberia's Yana River Valley and left behind numerous artifacts that enable anthropologists to get an idea of how they managed to live in the High Arctic, as described by Richard Stone and by V. V. Pitulko in the January 2, 2004, issue of *Science*. There were plentiful bone fragments of hunted animals (mammoth, bison, musk ox, horse, reindeer, wolf, fox, lion, and bear), as well as flaked stones, choppers, scrapers, and other tools, three of which were "foreshafts," used to replace the point of a spear quickly when it was broken or damaged. Two of the foreshafts were made of mammoth ivory, but one was of rhino horn, an 18-inch-long, gracefully curved shaft with a beveled point. Not only did early Europeans celebrate the great rhinos in their cave paintings, they used rhino horn as an aid in hunting big game—perhaps even rhinos.

The largest known prehistoric rhino was *Elasmotherium*, reaching a length of 20 feet and weighing as much as 4 tons—a rhino as big as an elephant. From nostrils to orbits, the skull was crowned with a large, bony dome, which served as a base for the single horn, leading paleontologists to reconstruct the horn of *Elasmotherium* as a very tall, very broad-based cone. This shape appears regularly in the Chinese sculptures from Gansu Province but has been found nowhere else. "Together with other archaic species of giant mammals, the unicorn *Elasmotherium* probably survived in the East Asian refugia until it was hunted to extinction in the late Pleistocene period," Parker speculates. There is no reason why a distant memory of this "unicorn" in Asia could not have survived into early Chinese history, contributing to the contention that the unicorn is indeed a Chinese invention.

The conical shape of the horn lent itself nicely for use as a drinking cup, and it was not long before the curative powers attributed to the powdered horn were transferred to the cup, meaning that a liquid drunk from a rhino-horn cup came to be seen as healthier than, say, a liquid drunk from a ceramic vessel. If there was poison in the drink, the rhino-horn cup would cause it to bubble and fizz. If the horns of *Elasmotherium* were not available for medicinal uses in early China, the horns of living rhinos at the time certainly were. During the Bronze Age (around 3000 BC), when China was warmer than it is today, all three types of Asian rhino were found in China: the Indian (or one-horned) rhino, the Sumatran, and the Javan rhino. The horns, dried and powdered, were (and still are) believed by the Chinese to be a "cold drug," a preventative against hot ailments, such as poisons and fevers.

The properties of "unicorn horn" were recognized in China long before they occurred to Europeans, probably because there were "unicorns" in China long before there were any in the West. In his fifty-volume pharmacological encyclopedia, the fourth-century Chinese author Li Chih-chen stated that the main ailments that could be treated with rhinoceros horn were snakebites, hallucinations, typhoid fever, headaches, boils, carbuncles, vomiting, food poisoning, and "devil possession," but in addition, "the unicorn horn is a safe guide to tell the presence of poison: when poisonous medicines of liquid form are stirred with a horn, a white foam will bubble up, and no other test is necessary."

Elasmotherium lived in north-central Europe and eastern China twenty thousand years ago. Is this gigantic fossil rhinoceros responsible for the myth of the unicorn?

(In their 2002 book *Rhinos*, Ann and Steve Toon suggested that, "improbable as it sounds, there may be some justification for the belief, as the alkaloids present in some poisons do react strongly with the keratin and gelatin in horn.")

Belief in the alexipharmic (poison-antidote) qualities of unicorn horn are as ancient as the earliest unicorn sightings. In *De natura animalium*, a miscellany of facts and fantasies gleaned from various Greek writers, the Roman writer known as Aelian (AD 170–230) had this to say about the horn of the unicorn:

> India produces horses with one horn, they say, and the same country fosters asses with a single horn. And from these horns they make drinking-vessels, and if anyone puts a deadly poison in them and a man drinks, the plot will do him no harm. For it seems that the horn both of the horse and of the ass is an antidote to the poison.

Some time between the second and the fourth centuries, the *Physiologus*, a book of animal legends, was produced in Alexandria and subsequently translated into Syrian, Arabic, Armenian, Ethiopian, Latin, German, French, Provençal, Icelandic, Italian, and Anglo-Saxon. This was not, however, a single work in serial translations but rather an ongoing work-in-progress with no single author, with material being added and modified as the work wandered through time and geography. The *Physiologus* eventually metamorphosed into the medieval bestiary. Since unicorns were one of the animals prominently featured in this important work, their place in history, mythology, and literature was assured. In his translation of a twelfth-century bestiary, T. H. White describes two unicorns, explaining, "The reason there are two unicorns described in this Bestiary is that Aelian believed there were two species—a solid-footed, donkey-sized one identified with the 'Indian Asse,' and a cloven-footed creature identified with the oryx." Of the *Monoceros*, this description appears in the *Physiologus* (as translated by T. H. White):

> A horn sticks out from the middle of its forehead with astonishing splendour to the distance of four feet, so sharp that whatever it charges is easily perforated by it. Not a single one has ever come

> alive into the hands of man, and, although it is possible to kill them, it is not possible to capture them.

Of the Unicorn:

> *Unicornis* the Unicorn, which is also called Rhinoceros by the Greeks, is of the following nature. He is a very small animal like a kid, excessively swift, with one horn in the middle of his forehead, and no hunter can catch him. But he can be trapped by the following stratagem. A virgin girl is led to where he lurks, and there she is sent off by herself into the wood. He soon leaps into her lap when he sees her, and embraces her, and hence he gets caught.

Through its appearance in bibles and bestiaries, we can trace the prominent tracks of the unicorn through the historical thickets of the Middle Ages. No less respectable an observer than Leonardo da Vinci wrote, "In its lack of moderation and restraint and the predilection it has for young girls, it completely forgets its shyness and wildness; it puts aside all distrust, goes up to the sitting girl and falls asleep in her lap. In this way hunters catch it." The actual "horns" that had begun to appear in Europe seemed ready validation for the convictions of those who believed in the fabulous unicorn. These "horns" were beautiful ivory shafts, wider at the base and tapering to a graceful point. What other animal could produce such a lovely object? In his 1560 *Historia Animalium*, Conrad Gesner had no problem; he saw the "horns" as incontrovertible evidence of the existence of the unicorn: "one has to trust the words of wanderers and far-going travelers, for the animal must be on earth, or else its horns would not exist."

In many early depictions of the unicorn, the animal is shown dipping its horn into a stream, ostensibly to purify the water of the poison that serpents have spewed into it. Several other illustrations included in Sutherland Lyall's *The Lady and the Unicorn* show the beast engaged in this unambiguous activity, and one from a fifteenth-century French manuscript shows the unicorn cleansing the water, while camels, leopards, lions, bears, and monkeys wait patiently until the water is pure enough to drink. In the second panel of the "Unicorn Tapestries," displayed in the Cloisters in New York, part of the Metropolitan Museum of Art, the animals also wait patiently to drink. Adolfo Cavallo sees the allegory in the European tradition as obvious: "Christ takes on the sins of Man, and so purifies him, in order to bring about his redemption. The serpent is the devil; the poison he introduces into the world (the water) is sin." If the unicorn used its horn for such eleemosynary purposes, it should come as no surprise that people—usually less charitable than unicorns—might use the horn in the same fashion, but to protect themselves, not their fellow creatures. Thus we read in Edward Topsell's 1658 *Historie of Foure-Footed Beasts* (with original spelling and punctuation retained):

> The ancient writers did attribute the force of healing to cups made of this horne, wine being drunke out of them: but because we cannot have cups, we drinke the substance of the horn, either by it selfe or with other medicines. I happily made some of this Sugar of the horne, as they call it, mingling with the same Amber, ivory dust, leaves of gold, Corall, and certain other things, the horne being included in silke, and beaten in the decoction of razens and Cinamon, I cast them in water, the rest of the reason of healing in the meantime not being neglected. It is moreover commended of Physitians of our time against the pestilent fever, against the bitings of ravenous Dogs, and the strokes or poysonsome stings of other creatures: and privately in rich mens houses against the belly or mawe worms; to conclude it is given against all poyson whatsoever, as against many grievous diseases.

Medieval Europeans, like their medieval Chinese counterparts, recognized the very special properties of some animal parts, but where the Chinese could attribute magical qualities to rhino horns, the Europeans had to make do with a mythological beast—until a real unicorn came along.

There are two sets of sixteenth-century "unicorn tapestries": one in the Cloisters and the other in the Musée de Cluny in Paris. In both sets, wherever the unicorn appears, his horn is a perfect narwhal tusk; long, white, tapered, and spirally twisted, which has suggested to some that there was some suspicion that the magical horn did not come from a terrestrial quadruped at all. Topsell quotes several authorities, one of whom says that "there be Birds in Ethiopia having one horn on their foreheads, and are therefore called Unicornus: and Albertus saith, there is a fish cald *Monoceros*, and hath also one horn." Topsell dismisses those who believe this nonsense as "vulgar sort of infidell people," but it is evident that the myth of the graceful, blue-eyed equine was threatened.

By the seventeenth century, a proper description of the narwhal had been obtained: it was (and still is) a smallish Arctic whale with a single elongated canine tooth that sticks straight out in front of its head to a length of eight feet. Only male narwhals grow this tooth; females are toothless. This tusk, unique in all the animal kingdom, was certainly known to Eskimos but was probably found by early Norse seafarers, and because it is solid ivory, it was used as a prow decoration for war galleys, used for sword hilts, and carved into small items like combs and chessmen. Brought to Europe by Elizabethan explorers and later by commercial whalers, the narwhal tusk contributed mightily to the myth of the unicorn, for it was much easier to believe that this elegant white horn came from a graceful hoofed animal than from a dumpy Arctic whale.

Even if it came from a dumpy little whale and not a snow-white quadruped, *Unicornum verum*—the true unicorn's horn—had some very special properties. A broadside published by a London doctor in the seventeenth century (reproduced in Beer's 1977 book *Unicorn*) specified the ailments that could be cured by partaking of drink that passes through the horn—a narwhal tusk that could be viewed by visitors to the doctor's house in Houndsditch:

> A Most Excellent Drink made with a true Unicorns Horn, which doth Effectively Cure these Diseases: Scurvy, Old Ulcers, Dropsie, Running Gout, Consumptions, Distillations, Coughs, Palpitations of the Heart, Fainting Fits, Convulsions, Kings Evil, Rickets in Children, Melancholly or Sadness, The Green Sickness, Obstructions, And all Distempers proceeding from a Cold Cause.

In addition to "drinking it warm at any time of the Day," the doctor has "prepared twelve Pils in a Box to be taken in three Doses, according to Directions therewith given, the Price is 2s. the Box."

In both sets of the unicorn tapestries—one in New York and one in Paris—the unicorn is shown with a tapering, spiral horn that closely resembles the tooth of a male narwhal.

It is unclear who first identified the narwhal as the progenitor of the unicorn. It could have been Olaus Magnus (1490–1558), the brother of the archbishop of Uppsala (who is remembered for his insistence upon the existence of sea serpents, drawing them into his map of 1539). He said that "the monoceros is a sea-monster that has in its brow a very large horn wherewith it can pierce and wreck vessels and destroy many men." Regardless of who first made the identification, the tooth of the narwhal was one of the most desirable objects of the Middle Ages, endowed with magical properties, collected by royalty, and worth more than its weight in gold. In his book *The Narwhal* (1993), Fred Bruemmer tells us that "the immensely lucrative trade in unicorn horns began in the late 18th century when a trader brought to Japan a collection of curios, among them a narwhal tusk from Greenland." Traders continued to bring narwhal tusks to Japan and China, where they were used as decorations and also ground up for medicinal or magical purposes. It appears that the tusks were so valuable that some whalers went north not to kill whales but only to trade for the ivory spires. The Danes, who bought them to sell in foreign countries, certainly had no reason to declare them to be fishes' teeth, and because no one could deny the corporeal existence of these splendid ivory shafts, the unicorn myth was kept alive by unscrupulous merchants.

During the eighteenth and nineteenth centuries, and into the twentieth, hundreds of tons of narwhal tusks found their way into the trading rooms of Europe and Asia, some from animals that had been harpooned by Arctic whalers, but mostly by way of barter with Eskimos. Because the bowhead whalers were eliminating the object of their fishery, they turned increasingly to ventures that would supply them with valuable narwhal ivory, as well as polar bear skins, sealskins, and walrus ivory, for which they traded tobacco, firearms, and ammunition. In 1912, some six hundred narwhal tusks, weighing approximately 3 tons, were taken out of just Pond Inlet by traders. In addition, closing ice conditions, known as *savssats*, trapped thousands of narwhals in the winter of 1914–15, and they were easily killed by hunters who simply stood by the holes and harpooned them as they surfaced to breathe. Randall Reeves and Edward Mitchell estimate that between 1915 and 1924, some eleven thousand narwhals were removed from the Baffin Bay and Davis Strait region, but the number of animals killed was undoubtedly higher.

Eskimos traditionally hunted narwhals for their own use by harpooning them and then killing them, resulting in few losses, but when high-powered rifles took the place of harpoons, many narwhals were wounded and dived under the ice to die uncollected. Females have no tusks and were killed as often as males, so the number of tusks collected represents only a proportion of the narwhals killed.

The tusk of the narwhal is but one of the elements in the myth of the unicorn. As we have seen, the unicorn has been represented in Chinese iconography for thousands of years. In Europe the myth predates the Vikings; Ctesias, Aristotle, and Pliny wrote about it before anyone had ever seen a narwhal tusk, and their works were well known in medieval Europe. Once the tusks began to appear, it was easy enough to fit the tusk to the fable. Only the narwhal possesses this tapered ivory shaft, but rhinoceroses, antelopes, and goats also contributed to unicorn mythology. In

A Greenland Eskimo narwhal hunt. The tapered ivory shaft that protrudes from the skull of the narwhal looks to our eyes as if it belongs on the forehead of a graceful quadruped rather than this dumpy Arctic whale.

Africa and Arabia there are several species of antelopes with straight black horns prominently ringed at the base, the oryxes and the gemsbok. An antelope with a pair of black horns does not look much like a unicorn with a single white horn, but the myth did not necessarily have to derive from actual sightings. From disparate sources, similar unicorn myths arose and persisted in Western and Eastern mythologies, and were somehow translated into pharmacological imperatives.

Were it not for their anomalous dentition, narwhals would probably have been ignored by everyone except the Greenland and Canadian Eskimos, who would have killed them—and still do—for their meat, sinews, and blubber, all of which they make good use of. But alas for *Monodon monoceros*, this innocent inhabitant of Arctic waters became the object of a concentrated hunt that goes on to this day, even without the attendant mythology of the horn's magical properties. Narwhal tusks, still traded by Eskimos, no longer command a king's ransom, but they are still valuable, perhaps because they are so scarce, but also because they are among the most beautiful of all natural objects. (One was sold in the 1980s at a New York auction house for $10,000.) The horn Martin Frobisher found on a 1577 voyage to Baffin Island was "reserved as a jewel by the Queen Majesty's Commandment, in her wardrobe of robes," and in the collection of Prince Takamatsu of Japan there are two "unicorn horns." At the Abbey of Saint Denis in Paris there is a 6-foot, 7-inch tusk that was said to have been presented to Charlemagne, and there is another in Strasbourg Cathedral.

The religious significance of the unicorn may have protected it, for despite the enormous value placed on its horn, there are no depictions of a unicorn hunt where the horn is removed from the dead animal. A unicorn could be captured with the assistance of a virgin, but what happens next was never explained. This suggests to Adolfo Cavallo (1998) "that people who believed in unicorns in the Middle Ages, and apparently most of them did, were not prepared to reduce it to the status of just another game animal." No such constraints affected those who hunted narwhals, the little whales that provided the horns that confirmed the medieval existence of the unicorn; and by the time rhinoceros horn was found to fulfill the same pharmacological functions, the hunters declared an all-out war on the rhinos and regarded them not even as "game animals" but as disposable impediments to the horn-collecting business.

There is little resemblance between a snow-white, delicate unicorn and a hulking, dirty-gray rhinoceros, but both sprout horns where few other animals have horns, and while the unicorn has proven to be more than a little elusive, rhinos are not that difficult to locate. Rhino horn is not ivory but rather compacted keratin fibers, the stuff of which hair and fingernails are made, so it would be just as efficacious to drink a potion made of powdered fingernails. It is not so much the composition of the substance that makes it magical, however, but rather its point of origin and its historical connection with the unicorn in Chinese and Western mythology. It was Marco Polo who introduced the rhinoceros to Europeans and helped make that connection.

Around 1265, two Venetian brothers named Niccoló and Maffeo Polo became the first European merchants to reach the Mongol court of Kublai Khan in China. They returned to Europe in 1269 and set out again for the East two years later, accompanied by Marco, Niccoló's teenaged son. Marco Polo (1254–1324) remained in China until 1292 and then returned to Venice, where he joined the Venetian forces fighting Genoa and was taken prisoner. During his two-year incarceration (1296–98), he dictated his memoirs to a Pisan writer named Rustichello, who transcribed the traveler's amazing adventures during seventeen years at the Mongol court, including visits to much of Asia and the Arab world. There are some who doubt the accuracy of some of Marco Polo's accounts, but his description of the "unicorn" rings true, for no other reason than it is a fair description of a Sumatran rhinoceros, likely the first ever seen by a European. Here is Marco Polo's account:

> They have wild elephants and plenty of unicorns, which are scarcely smaller than elephants. They have the hair of a buffalo and feet like an elephant's. They have a single large, black horn in the middle of the forehead. They do not attack with their horn, but only with their tongue and their knees; for their tongues are furnished with long, sharp spines, so that when they want to do any harm to anyone they first crush him by kneeling upon him and then lacerate him with their tongues. They have a head like a wild boar's

Although it is made of compacted hair fibers, rhinoceros horn can be carved with intricate detail. This example, from the Qing Dynasty of eighteenth-century China, shows that the horn could be used for purposes other than medicinal.

> and always carry it stooped towards the ground. They spend their time by preference wallowing in mud and slime. They are very ugly brutes to look at. They are not at all such as we describe them when we relate that they let themselves be captured by virgins, but contrary to our notions.

Marco Polo described the Sumatran rhinoceros as a unicorn, but not the kind that could be captured by a virgin. (Indeed, while a somnolent unicorn might be carried away, it is more than a little difficult to imagine how one might haul away a snoozing rhino.) Nevertheless, the dainty unicorn, which never actually existed, and the ponderous rhinoceros, which exists today (but only barely), have been pharmacologically conjoined for at least eight centuries.

Willy Ley (1906–69), the Berlin-born scientist and a prolific author on such subjects as space travel and romantic zoology, was particularly interested in the transmogrification of the ethereal unicorn into the more prosaic rhinoceros, and in *The Lungfish, the Dodo, and the Unicorn*, he wrote:

> Rhinoceros horn might be passed off for "true" alicorn [he used the term *alicorn* to avoid the cacophonous "unicorn horn"]. . . . In Northern Europe *unicornum verum* and *unicornum falsum* were strictly distinguished. The former "true alicorn" was usually found in the earth—actually mammoth tusks—while the false alicorn came from the north—actually the tusk of the narwhal.

By the eighteenth century, as alicorn was disappearing from European pharmacies, it was beginning to appear on the shelves of their Chinese equivalents, as it was an important element in Chinese medicine. Unlike the horn of the unicorn, the horn of the rhinoceros was relatively easy to collect; the animal now known to science as *Rhinoceros unicornis* used to be found throughout the northern Indian subcontinent. There are rhinos clearly depicted on Harappan sealstones from Mohenjo Daro (now in southern Pakistan) that have been dated as far back as 2000 BC. Curiously, in the Indian culture whose religious iconography includes elephants, monkeys, cattle, and snakes, the rhinoceros, a large and visible creature whose nose-horned visage suggests mystery, appears mostly in hunting scenes. The same is true of tigers, but tigers were a threat to humans, while rhinos, for the most part, were not.

Indeed, the opposite was true: poachers turned out to be the greatest threat to the world's rhinos, and there was nowhere on earth that the great beasts were safe from the predation of humans, who now hunt them primarily for their horns. If you thought that modern medicine has no place for something as archaic as rhinoceros horn, you would be sadly, tragically, wrong.

The Nose-Horns

Of the five living species of rhinoceros, three in Asia and two in Africa, all are threatened by the relentless demands of traditional Chinese medicine. The Asian rhinos are the great Indian (*Rhinoceros unicornis*), the

Taken from Albrecht Dürer's famous drawing of 1515, Conrad Gesner's rhinoceros was published in his 1551 *Historia Animalium.* Like Dürer's rhino, Gesner's also has a tiny horn protruding from its withers, where no proper rhino has a horn.

Javan (*Rhinoceros sondaicus*), and the Sumatran (*Dicerorhinus sumatrensis*). African rhinos are the white rhino (*Ceratotherium simum*) and the black (*Diceros bicornis*). The Indian and Javan rhinos have a single horn on the nose, while the Sumatran and both African species have two. (The scientific names of the various rhino species are helpful in determining the number of horns on the nose: *Rhinoceros* means "nose-horn," *unicornis* means "one horn," and *bicornis* means "two horns.")

Male African rhinos often use their horns for fighting, and in David Macdonald's *Encyclopedia of Mammals*, we read that "black rhinos have the highest incidence in mammals of fatal intraspecies fighting: almost 50 percent of males and 33 percent of females die from wounds." Like African rhinos, Indian rhinos have no natural enemies in the wild, but they use their teeth rather than their horns to fight with each other, according to Stanley Breeden and Belinda Wright in *Through the Tiger's Eyes*. All rhinos have a massive body, a large head, and short stumpy legs. Their eyesight is poor, but their hearing is acute and their sense of smell is excellent. Like the elephants with which they share most of their habitats, rhinos are "megaherbivores" and require prodigious amounts of food to support their great bulk.

The horn of each species, as already mentioned, is composed not of ivory but of fibrous keratin, the same stuff as your hair and fingernails. But it would be wrong to see the horn as simply a mass of densely packed hair. Toward the tip, rhino horn can be worn smooth and shiny; held up to the light, it has a translucent amber glow, which makes it a highly desirable material for carvings. Ground into powder, rhino horn is one of the most valuable substances in the world. Today, the Javan and Sumatran rhinos are on the brink of extinction, and the Indian, and the African (black and white) rhinos are in serious trouble, thanks in recent decades almost entirely to the desire for those horns, mainly, but not exclusively, for use in traditional Chinese medicine. The pressure on each of the five rhino species is somewhat different.

The Rhinos of Africa

The Black Rhino

"The black rhino," wrote Carol Cunningham and Joel Berger in 1997, "is nature's tank. A charging rhino can gallop at 50 kilometers per hour, chug through dense thornbush, and scatter a herd of elephants. . . . Stalking lions will break off a hunt to detour around them. . . . At the turn of the century, there may have been 100,000 in Africa, scattered from below the Sahara to the Cape. Now in the entire continent, only one unfenced population of more than 100 animals remains."

Not actually black, but brownish gray, the black rhino grows to a length of 12 feet, stands close to 5 feet high at the shoulder, and can weigh almost 2 tons. It is recognizable by its long, pointed, prehensile upper lip and two prominent horns, the longest of which averages 20 inches, but in some animals the forward horn may grow to be much longer. The longest black rhino horn on record—nearly 5 feet in length—belonged to a female nicknamed Gertie, who lived in the Amboseli Reserve in Kenya in the 1950s and was a darling of photographers. Her forward horn, as described by Guggisberg in 1966, "was certainly not under forty inches, probably quite a bit longer, inclined forward at an acute angle, and showing a slight, but graceful curve at the tip." C. A. Spinage (1962) put together the evidence for its length this way:

> From a study of photographs taken in 1952 Gertie's horn must have then been around forty inches long. The major portion which was recently broken off was recovered and found to be thirty-nine and a half inches in length; this was matched against photographic enlargements of the intact horn in life and its total length was thus estimated at fifty-four and a quarter inches. So it would appear to have grown about eighteen inches in six to seven years.

The black rhino inhabits the acacia scrub that grows at the edge of the plains where it browses off the ground on a wide variety of plants, especially regenerating twigs, which it gathers into its mouth with the prehensile upper lip. The two horns typical of both black and white rhinos grow continuously throughout the animal's life; horns that are broken off can regrow. Because of their notoriously poor eyesight, black rhinos sometimes charge a disturbing sound or smell, and they have been known to toss people into the air and upend cars.

Major A. Radclyffe Dugmore, an American wildlife photographer, painter, and printmaker who turned from hunting to capturing his sub-

African Black Rhino (*Diceros bicornis*).

The longest measured forward horn of a black rhino was 44 inches in length. At their current rate of extirpation, it will not be long before the only way to see black rhinos in a "natural" setting will be in a museum diorama, such as this one in the American Museum of Natural History in New York.

jects on paper, saw and photographed East African wildlife in 1909–10, and in *The Wonderland of Big Game*, published in 1925, he remarked on the decline even then in the black rhino population:

> The most notable decrease among the animals is that of the poor old rhino, notwithstanding what anybody may say to the contrary, and some observers may challenge my statement. During my first visit to Kenya I saw as many as thirteen in sight at one time, and groups of four or five were not uncommon. During my last trip, when I covered a large area of country and visited many places where rhino used to abound I saw thirteen altogether. The ease with which the stupid creatures may be shot must account for this, coupled of course with the idea, prevalent with many people, that it is a noteworthy feat to kill the wretched brute. Unless very stringent laws are made for their protection, it is safe to predict their early

extermination, except possibly in forest country, where they still live more or less unmolested.

In his 1966 book *S.O.S. Rhino*, C. A. S. Guggisberg blamed the decline in rhino numbers not on its stupidity, but on ours: "In the case of the rhino the illogicality and stupidity of mankind has resulted in a situation in which the world is in danger of losing the improbable but altogether fascinating rhinoceros forever." Guggisberg summarized the status of black rhinos in various regions of Africa:

> In 1960 the number of black rhinos surviving in Kenya was estimated at 2,500. . . . As far as Uganda is concerned, a few black rhinos lived in the Masaka and Ankole Districts within the memory of early residents. . . . By 1957 the rhinos of Northern Acholi had been practically exterminated by poachers coming over from the Sudan, and the stronghold of the species was definitely in the Kidepo Valley of North Karamoja. . . . In the eastern Sudan . . . the animals had become rare by 1912, and could only be found much farther to the south, along the Dinder River. . . . Moving westwards we find the same dismal story everywhere. Incredible numbers must have been butchered during the last fifty years in all the countries lying between the Sudan and Northern Nigeria. . . . Rhinos have become scarce in Central and North East Cameroon, very rare in the Lake Chad area and may be extinct in Nigeria. . . . On the Ivory Coast the last individuals were shot near Bouna in 1905. . . . No European is known to have seen a rhinoceros in the former French territory of Niger.

At one time, black rhinos were considered premier big-game trophy animals, along with the other four creatures that made up the "big five": lion, leopard, buffalo, and elephant. In *Green Hills of Africa*, Ernest Hemingway described a black rhino he has just shot: "There he was, long-hulked, heavy-sided, prehistoric-looking, the hide like vulcanized rubber and faintly transparent looking, scarred with a badly healed horn wound that the birds had pecked at, his tail thick, round, and pointed, flat, many-legged ticks crawling on him, his ears fringed with hair, tiny pig eyes, moss growing at the base of his horn that grew outward from his nose. . . . This was a hell of an animal."

People have always killed rhinos, but not always for sport or even for medicinal purposes. In his 1952 book *Hunter*, the Scotsman J. A. Hunter claimed to have killed a thousand rhinos, but the actual number is probably much higher. In the chapter entitled "The Great Makueni Rhino Hunt," he explained that he was hired to kill the rhinos because the tsetse flies made it impossible to raise cattle, and the only way to eliminate the flies was first to clear the bush of rhinos so that the labor gangs could safely work there. Peter Beard's *The End of the Game* contains a reproduction of Hunter's notebook page for Makueni, in which we see that he actually killed 165 rhinos between August 29 and November 28, 1944; from June 29 to December 31, 1945, he killed 221 more; and from January 1 to October 31, 1946, he killed another 610, for a total of 996. (The land proved to be too poor for agriculture, so 996 rhinos died for nothing.) If he killed 996 rhinos just between August 1944 and October 1946 in what he called "the biggest rhino hunt in history," then he must have killed far more than a thousand, for he was hunting for more than forty years.

According to Clifton Fadiman's commentary in the Book-of-the-Month Club brochure advertising the book's initial publication, the killing didn't stop there. Hunter had "joined the Kenya Game Department as a control officer, charged with the extinction of animals that were endangering native holdings or upsetting the region's ecological balance by becoming too numerous for their environment. In the course of his duties he has shot more than 1,400 elephants, and is said to hold the world's record for rhinos and possibly lions."

Because their scrubland habitat has remained largely unoccupied by settlers and has not been much in demand for agriculture—except occasionally by the Kenya Game Department—black rhinos were not seriously threatened until the 1970s, when the demand for horn began in earnest, first for dagger handles and then for the larger market that was traditional Chinese medicine. As Guggisberg wrote, "Up to that time, the horns of the two African species had mainly served for the carving of drinking cups, snuff boxes, handles for knives, swords and war-axes, no superstitious significance whatever being attached to them by any of the indigenous people of Africa." When it was discovered that rhino horn could be sold for huge amounts of money to suppliers of Asian traditional medicines, the onslaught began: no African rhino was safe.

Now there are so few rhinos left in parts of Africa that many are literally kept under armed guard. They forage during the day, accompanied by guards with rifles, and they are locked up at night. Yet rhino horn is so valuable that poachers have killed guards to get at the animals. "Around 1900, the savannas and woodlands of sub-Saharan Africa may have supported more than 1 million black rhinos," Eric Dinerstein (2003) suggests, but recently, their numbers have been in free fall. Through the pages of the journal *Pachyderm* (the journal of the African Elephant, African Rhino, and Asian Rhino Specialist Groups of the IUCN), we can track the precipitous decline of the black rhino population. David Western and Lucy Vigne (1984) estimated that in 1984 there were 8,000–9,000 black rhinos left. In 1987 Esmond Bradley Martin (1987b) wrote, "One must remember that in 1970 there were 65,000 and by 1980 only 15,000 were left." Western (1989) wrote that "since 1970 its numbers throughout Africa have declined . . . to around 3,800 in 1987."

There are now about 3,600 black rhinos in all of Africa, in small pockets in Zimbabwe, South Africa, Kenya, Namibia, and Tanzania. "The black rhinoceros," noted Milliken, Nowell, and Thomsen in 1993, "has declined at a faster rate than any other large land mammal in recent times, making a rapid transition from abundant to endangered." It is impossible to exaggerate the magnitude or the significance of the collapse of the black rhino population. In the definitive *Mammals of the World*, Ronald Nowak (1991) called it "the greatest single mammalian conservation failure of the twentieth century."

Living in Africa has provided no special protection for the black rhino from Asian pharmacological prescribers, especially since the stock of Asian rhinos was greatly depleted; its horns are collected by African poachers and sold to dealers to be converted to traditional medicines in Asia. In addition, because young men in the Arab country of Yemen covet rhino horn for the handles of their elaborately carved daggers called *jambiyas*, another major threat to the already beleaguered African rhinos began to arise when Yemeni incomes rose. When a Yemeni boy approaches manhood, he is given his own dagger, more as a weapon than as an ornament. Especially in and around the capital city of Sanaa, nothing is thought to make a better jambiya handle than the horn of a rhino; it takes a high polish and improves with age. Before about 1970, few men could afford these status symbols, but during the oil boom of the 1980s, many Yemeni men traveled north to work in the lucrative Saudi oil fields, where they were able to earn more money in a year than they had earned in their entire lives, so they were now able to buy the heretofore prohibitively expensive jambiyas. There was a sevenfold increase in the per capita income in Yemen during that time and a *twentyfold* increase in the price of rhino horn.

The increased price of rhino horn made it enormously profitable to poach rhinos and sell the horns on the black market, and Yemen suddenly became the world's largest importer of rhino horn. East African customs records show that from 1970 to 1976, 634 kilograms (1,394 pounds) were exported each year to Yemen, while North Yemen's official statistics show 2,878 kilograms (6,331 pounds) being imported annually, according to a 1997 article written by two indefatigable investigators of the plight of the rhinos, Esmond Bradley Martin and Lucy Vigne. In 1990, they tell us, the two horns from a single black rhino brought $50,000, while in 1992, the Sheikh of the Bakils, Yemen's largest and most powerful tribe, paid a million dollars for a jambiya that had been owned by a previous North Yemen ruler. A collapse in oil prices in the mid-1980s, along with a government ban on the import of rhinoceros horn, resulted in the official decline of the dagger-handle trade, but substantial amounts are still being smuggled in.

Like poaching for elephant ivory, poaching for rhino horn is simply too profitable for many subsistence farmers and herders to resist. In a 1985 article on Yemeni dagger handles, E. B. Martin wrote, "During the past 15 years, the decline of the rhino has probably been more acute than that of any other large mammal in Asia or Africa. It is almost ironic that international conservation organizations have, in the same period, spent very large sums to help control the ivory trade. There are at least 750,000 elephants in Africa, as compared to 12,500 rhinos" (1985b). (That was in 1985; as of 2005, there were 3,600 black rhinos left.)

After a decline in the 1990s, for reasons not clearly understood, the popularity of rhino-horn dagger handles took another upturn in 2000.

A Yemeni man proudly displays his jambiya with a rhino-horn handle in the market at Sanaa, where most jambiyas are made. The most expensive one ever recorded sold for a million dollars in the early 1990s.

In *Pachyderm*, Martin and Vigne noted that the price of rhino horn in Yemen had increased once again (2003). The Environmental News Service (http://ens-news.com) quotes Russel Taylor of the World Wildlife Fund (WWF) office in Harare, Zimbabwe, as saying in 2003 that "rhinoceros horn is the most highly priced commodity in the world." From 1998 to 2002, a total of forty-six rhinos were killed in the Democratic Republic of the Congo, Kenya, and Tanzania, while poaching increased in East Africa as well, resulting in more rhino horn sold in Yemen, even though many of the shops were offering jambiyas with water buffalo-horn handles alongside those with the traditional rhino-horn handles.

All worldwide trade in rhino horn is now prohibited, because rhinos are protected under Appendix 1 of CITES (the Convention on International Trade in Endangered Species of Wild Flora and Fauna) but the ban has not been very successful because of the thriving black market. In 1993, the United States threatened to ban legal imports of wildlife from China, which has a large wildlife trade with the United States, if China did not start taking measures to stop illegal wildlife trade. In response, China made it illegal to sell, buy, trade, or transport rhino horns and tiger bones. Illegal stockpiles of rhino horns and tiger bones remain, however, and can be sold for astonishing prices.

Despite the Chinese ban, however, the trade in rhino horn continues. Because a jambiya handle is not shaped at all like a rhino horn, there is a lot of waste in the carving process. "Wastage in making rhino horn handles for jambiyas runs to over 60%," wrote Martin and Vigne in 2003. "Large quantities of rhino horn chips and powder since the 1970s have been sent from Yemen to Chinese pharmaceutical factories." Nothing is wasted in this process—except of course, the rhino.

The White Rhino

Whether the African rhino was white or black (or any color in between), its horns were required for dagger manufacture and the practice of TCM. White rhino horns are just as desirable for jambiyas as black; they are both shaved and powdered for traditional Asian medicines. Like those of other species, white rhino horns can be hollowed out, polished and made into cups, or carved into small statues, buttons, hairpins, combs, and walking-stick handles; the feet have been made into the (mercifully obsolete) ashtrays and umbrella stands.

The white rhino is no more white than the black rhino is black. Its name has long been held to be a corruption of the Dutch word *wijdt*, which means "wide," and refers to its broad muzzle. Standing more than 6 feet high at the shoulder and weighing as much as 3 tons, the white rhino is the largest living land animal after the African and Indian elephants. Like the black rhino, the white rhino has two horns, and the front one is usually much longer than the rear one. Also like the black rhino, the record length for the front horn is a little less than 5 feet. Unlike the black rhino, however, the white has a pronounced hump between the shoulders and a long, squarish head, which is so disproportionately large that the animal appears unable to hold it up, and it is often seen with the muzzle close to the ground. An animal of open grasslands, *Ceratotherium simum* was once widely distributed throughout sub-Saharan Africa, wherever there was suitable savanna country. White rhinos are true grazers, feeding on short grasses that grow close to the ground. Like other rhino species, the white was shoved off its land by people who wanted to raise cattle on the plains and was slaughtered for its horns.

There are two recognized subspecies of white rhinoceros, the northern (*Ceratotherium simum cottoni*) and a southern (*C. s. simum*); they have been shown to be genetically distinct, and where the northern subspecies is nearly extinct, the southern still thrives. (It is the only rhinoceros—species or sub—that is not critically endangered.) The northern subspecies was once found in southern Chad, the eastern

African White Rhino (*Ceratotherium simum*).

Central African Republic, Sudan, and the Democratic Republic of the Congo. In 1960, there were more than one thousand northern rhinos in Garamba National Park, but in 1963, rebel forces entered the park from Sudan, occupied virtually the entire park, and killed more than nine hundred of them, according to Fisher, Simon, and Vincent, in their 1969 book, *Wildlife in Danger*. More recently, Sudanese marauders known as "horsemen" have killed almost one thousand elephants and are on the verge of eliminating the last of the northern white rhinos. The heavily armed poachers use pack animals to carry out the ivory tusks and rhino horns, and leave the entire carcass behind. The gravity of the situation was pointed out by Andrew Revkin in a *New York Times* article of August 7, 2004:

> The continent's last known wild population of northern white rhinoceroses has been halved by poaching in the last 14 months, according to aerial surveys done in July of Garamba National Park in Congo. The survey estimated that 14 to 19 rhinos were shot in the park, cutting the overall population despite the birth of four calves this year, said scientists of the World Conservation Union (IUCN). . . . Armed poachers who slaughter them smuggle the horns to Asia, where they are coveted for purported medicinal properties.

In the months that followed, suggestions were made to airlift some of Garamba's rhinos to a safe haven in Kenya. Quoted in a Reuters report by David Lewis, rhino specialist Kes Hillman Smith said it was "the safest means of securing the subspecies from extinction." But after authorizing the plan, the Congolese government reversed its field and decided that the issue was one of "national sovereignty and it should look after its own" (Anon 2005). The remaining rhinos of Garamba are stuck in the middle of a human territorial dispute, and even if the last rhino is not killed by poachers, this unfortunate incident demonstrates how precarious a population can be if its numbers are greatly reduced.

The southern subspecies has fared much better. Formerly found throughout southeastern Angola, central Mozambique, Zimbabwe, Botswana, eastern Namibia, and northern South Africa, the southern white rhino was the target of an organized hunt in the nineteenth century and was nearly eliminated. Unlike the often belligerent black rhino, the white is a docile creature, and hunters were able to walk right up to some of them and shoot them where they stood. Others were trapped in spike-floored pits, while still others were speared from horseback. Whatever the method, thousands of white rhinos were killed, and their horns hacked off and shipped to Asia. Most of Africa's white rhinos are now confined to protected areas and game farms, and there has been a resurgence in their numbers. There are well over ten thousand white rhinos in the south, and they are no longer considered endangered. In addition, there are more than seven hundred southern white rhinos in zoological parks around the world. (For some reason, northern white rhinos do not do well in captivity, and as of 2002, there were only nine individuals in two zoos, one in the Czech Republic and the other in San Diego.) The white rhino population is larger today than that of all other rhino species combined, and the rescue of the southern white rhino is the only success in the largely negative history of the modern interaction of rhinos and men.

The Rhinos of Asia

> Although modern rhinos are far more restricted in distribution and diversity than was the group in the geological past, it would be wrong to think that they are inevitably doomed to a natural extinction. Even in the nineteenth century they occurred in large numbers over much of Africa and Asia. The subsequent population crashes have been entirely the fault of relentless killing and habitat usurpation by people. Nearly all parts of rhinoceroses are used in folk medicine, but by far the greatest demand is for the horn, which in powdered form is reputed to cure numerous physical problems and which whole is used for artistic carving.
>
> *Ronald Nowak, Walker's Mammals of the World (1991)*

The Sumatran Rhino

"Although modern rhinos are far more restricted in distribution and diversity than was the group in the geological past," wrote Ronald Nowak in *Walker's Mammals of the World*, "it would be wrong to think that they are inevitably doomed to a natural extinction. . . . The subsequent population crashes have been entirely the fault of relentless killing and habitat usurpation by people. Nearly all parts of rhinoceroses are used in folk medicine, but by far the greatest demand is for the horn, which in powdered form is reputed to cure numerous physical problems and which whole is used for artistic carving." It is the fate of the three Asian rhino species to live in the very regions where their horns are considered essential elements in traditional medicine—and more valuable than gold.

In his 1929 *Field-Book of a Jungle-Wallah*, British naturalist Charles Hose (1863–1929) wrote of his earlier adventures in Sarawak:

> The Borneo [Sumatran] rhinoceros is a smallish species, and quite the most grotesque of his kind: he has two horns and his hair is rough and bristly, almost like fine wire. He frequents the foot-hills below the mountains, and comes down in the heat of the day to take his ease in what are called "salt-licks," muddy baths formed by springs of saltish water. The clearing in the mud and the bushes were, I was told, caused by the creature's trampling movements to his lair higher up the hills.

Later, G.T.C. Metcalfe summed up the status of the Sumatran rhino, known in Malaysia as *badak* in this way: "Persistent persecution of this rhinoceros in the past has driven it into the most inaccessible, uninhabited, and usually hilly tracts, with the result that it is extremely difficult to obtain accurate information as to its whereabouts." In what must be seen as an attempt to protect this bedeviled creature from further poaching, he added:

> The supposed medicinal value of any part of a rhinoceros is yet to be substantiated and a number of other beliefs relating to the value of the horn as an aphrodisiac and to its nullifying effect on poisons is entirely erroneous. . . . Since the destruction of the rapidly declining rhinoceros population is entirely to obtain parts for their supposed medicinal properties, a ban on the import, export, and possession of all rhinoceros parts, especially the horn, which cannot be positively identified and proved to be covered by a specific certificate for the Game Depart-

Sumatran Rhino (*Dicerorhinus sumatrensis*).

ment (or its equivalent) in the country of origin, would do much to stop the illegal killing of animals, not only in Malaya but elsewhere.

After 1963, "Malaya" became Malaysia, Singapore broke away to become a separate nation in 1965, and a ban on the import, export, or possession of rhino parts was passed. But because this species of rhino has the misfortune to live in some of the most densely populated countries in the world, where a large proportion of the populace believes that rhino horn has powerful medicinal properties, its future is grim indeed.

The Sumatran rhino (*Dicerorhinus sumatrensis*), also known as the hairy rhino because of its furry coat, is the smallest of the living rhinos, weighing a maximum of 1,000 kilograms (about a ton), compared with the Indian or white rhinos, which can weigh more than twice as much. This species has two horns, but the forward one is sometimes so small as to be inconspicuous. Sumatran rhinos prefer dense rainforests and are usually found close to water; their diet consists of shoots, twigs, young foliage, and fallen fruit. Originally found in India and Bangladesh, south to the Malay Peninsula and Vietnam, they are now found in low numbers on the island from which they take their common name; there may also be twenty to thirty in the deep rainforests of Indonesian Borneo (Kalimantan), and another six or seven animals in Burma (Myanmar).

Even with animals as large as rhinoceroses, it is not easy to count them in dense jungles, and it has been estimated that there are now fewer than three hundred Sumatran rhinos living anywhere in the world, and those in very small and highly fragmented populations, with little improvement in sight.

In historic times, the hairy rhino was found throughout north-eastern India, but the last two records of its presence anywhere on the Indian subcontinent were in 1967, when, in the words of Anwarud-din Choudhury (1997), "a Sumatran rhino was killed near Cox's Bazaar in the Chittagong area, and a [Sumatran] rhino was seen by local people in the Punikhal area of Sonai . . . southern Assam." In Way Kambas National Park, at the southern tip of Sumatra, what was thought to be the last Sumatran rhino there was shot in 1961, but in a survey conducted from 1991 to 1994, Joanne Reilly, Guy Spedding, and Apriawan indicated that a few of these elusive animals might still be found in the lowland rainforests of the park. In 1996, Dwiatmo Sis-womartono and colleagues published a color photograph of a Suma-tran rhino taken with an infrared camera trap in the park, unequivocally confirming the occurrence of at least one rhino in this Sumatran preserve.

In Myanmar, when Alan Rabinowitz, George Schaller, and U Uga surveyed the Tamanthi Wildlife Sanctuary in 1994, they found evi-dence of twenty-one species of medium-to-large mammals, including elephants, tigers, leopards, gaurs, bears, sambar, and barking deer—but no sign of the Sumatran rhino. Sightings of the species had been reported up to the 1980s, but after that there were only rumors, they learned.

In response to the calamitous decline of the Asian rhinos, the Asian Rhino Specialist Group was created in the 1980s by the Species Sur-vival Commission of the IUCN. From this came a plan to capture the "doomed" rhinos and remove them to various breeding facilities in Asia, Europe, and North America. This was done, with some disastrous results. The Sumatran Rhino Trust (SRT) was established to facilitate the export of animals for the foreign breeding programs, but protests over shipping Sumatran rhinos to Western zoos resulted in the dissolu-tion of the proposed agreement and the establishment of two separate programs in Malaysia. A new agreement was reached, and the rhinos were captured for transportation. In a June 1995 report in *Conservation Biology*, tellingly entitled "Helping a Species Go Extinct," Alan Rabinowitz tore into the plan:

> In 1993, the SRT was dissolved after five years and a cost of more than US$2.5 million. Virtually none of the money went to improving the protection and management of wild rhinos in existing protected areas. This program, along with the similar efforts in Sabah and Peninsular Malaysia to catch doomed rhinos for breeding, were expensive failures resulting in the capture of 35 rhinos and the deaths of 12 rhinos between 1984 and 1993. The failure was partly a result of the skewed sex ratio of captured animals. Still, as of 1993, the surviving 23 rhinos (14 females, 9 males) were being held in 10 separate areas in Indonesia, Peninsular Malaysia, Sabah, the United Kingdom, and the United States. Other than one facility in Peninsular Malaysia with five rhinos, no more than three rhinos were at any of the other facilities. Because adult males and females were never together in the same place for a significant amount of time, there have been no births from captive Sumatran rhinos to date, except one female who was pregnant when captured.

Rabinowitz concluded his report with these words: "Meanwhile, the decline of the Sumatran rhino continues. In August 1994, 12 more Sumatran rhino horns were confiscated in Taiwan that had been smuggled on a fishing boat from Malaysia (The *Jakarta Post*, August 9, 1994). After all these years, do we know how many Sumatran rhinos we are dealing with? No, but soon we might have a nice round figure."

Not surprisingly, Rabinowitz's scathing condemnation brought out a flurry of responses, both positive and negative, printed as Letters to the Editor in the journal. Among the criticisms, three IUCN biologists said that Rabinowitz had committed "several serious errors of commission and omission"; the Director of Conservation for the Indonesia Ministry of Forests complained that the author had made "a number of allegations in relation to Indonesia that misrepresent the current situation"; and an official of the Sabah Wildlife Department loyally defended the actions of his department. The last word belongs to Rabinowitz, who finishes his letter with this:

> Frankly, there is only one fact that is important right now—the Sumatran rhino is in desperate trouble and, to the best of our knowledge, is still sliding toward extinction. Good intentions and the most articulate of excuses mean nothing in this light. If this species is lost, after having watched its decline for so long, we have only ourselves to blame.

In his 2003 book about Asian rhinos, Eric Dinerstein also criticizes what he refers to as the "disastrous captive-breeding attempt financed by the Sumatran Rhino Trust, which left many rhinos dead in breeding stations. Even worse, recruitment in captivity was zero until the calf was born at the Cincinnati Zoo in 2001." When that 72-pound Sumatran rhino calf was born, it was an event of such importance in the annals of rhino preservation that its photograph appeared in color on the cover of the 2001 issue of *Pachyderm.** Mohid Khan, Thomas Foose, and Nico van Strien, the authors of the accompanying journal article, commented: "The species is neither safe in the forest, nor, more figuratively, yet out of the woods. A single birth is a significant breakthrough but in itself does not ensure the survival of the species." Tajuddin Abdullah, however, who was instrumental in starting Malaysia's rhino conservation program, now believes that sending rhinos to a captive center is tantamount to killing them. "We now hold the world's worst record for rhino conservation," he said in a December 2003 interview for the *Malaysia Star*. "If we capture more rhinos in the wild just to continue the breeding programme, we are just sending this rare species to their death cell." (By 2004, all Sumatran rhinos held in captivity by the Peninsular Malaysian Wildlife Department had died of disease. It is rumored that the department plans to capture more from the wild to replace those that died, maybe in a new facility, but this would be an invitation to repeat disaster.) The debacle

* On July 30, 2004, the Cincinnati Zoo announced that Emi, the female Sumatran rhinoceros, became the first Sumatran rhino in history to produce two calves in captivity. In her indoor stall, she delivered her second, a healthy female calf, at 12:51 p.m. "This is a historic birth. It is proof that the science of breeding Sumatran rhinos has been developed at the Cincinnati Zoo and the first birth was not a one-time wonder," said Dr. Terri Roth, Vice President of Animal Sciences. "Because Sumatran rhinos are on the brink of extinction, this calf serves as a lifeline for a species clinging desperately to survival" (*Cincinnati Zoo Newsletter*).

Female Sumatran rhino "Emi" with her second calf, "Suci" (pronounced "sue-chee"), born on July 30, 2004, at the Cincinnati Zoo. The breeding program at the Ohio zoo is more successful than any that have been tried in Asia.

with the Sumatran rhino project made many conservationists rethink the idea of preserving rhinos by distributing small numbers to various widely spaced zoos; it appears a much better idea to leave them in their own habitat, and, somehow, prevent people from killing them.

The Sumatran rhino, *Dicerorhinus*, according to Colin Groves and Fred Kurt (1972), "is the genus that gave rise to all living Rhinocerotidae; in that sense it can be considered a 'living fossil.'" In *The Future of Life* (2002), E. O. Wilson chose the Sumatran rhino as one of the paradigms for the current worldwide extinction crisis:

> Can the Sumatran rhino, like the California condor and the Mauritius kestrel, be pulled back from the grave? Of the two standard methods, captive breeding has been so far unproductive, while protection against poachers in existing reserves seems tenuous at best. The small circle of rhino experts working on the problem agree that *Dicerorhinus sumatrensis* has entered the endgame. Whatever the solution, they say, it must be found now or never.

Eric Dinerstein has found that the other rhino species, in many instances, are not faring so well either:

> The course of conservation, particularly in Thailand, Laos, Cambodia, and Vietnam, is quite different from what I have experienced on the Indian subcontinent. In Indochina wildlife reserves exist in name only; intense poaching has decimated the large mammal populations of this region. One is lucky to encounter tracks of large vertebrates in protected areas, let alone be blessed with an actual sighting. Why is this so? In stark contrast to the Indian subcontinent, Southeast Asia has no tradition of strict protection within nature reserves. But without strict protection, large mammals continue to disappear.

The Sumatran rhino is so close to extinction that the removal of a single animal brings them that much closer to the edge. They were brought to this precipice by poachers who hunted them for their horns and by loggers who destroyed their habitat. An October 2003 report that Myanmar's hardwood forests are being liquidated by Chinese timber companies from neighboring Yunnan Province does not bode well for the remaining rhinos there. Will the Sumatran rhino become the first animal to disappear completely because of traditional Chinese medicine?

The Javan Rhino

The Javan rhino (*Rhinoceros sondaicus*), the other rhino species found in Indonesia (with a remnant population in Vietnam), is somewhat larger than the Sumatran rhino. Sometimes known as the lesser Indian rhino because of its proportionally smaller head, no "rivets" on its leathery skin, and a slightly different pattern of folds making up its body armor, the Javan rhino is among the rarest large mammal species in the world. With no more than sixty individuals in the wild and none in captivity, it is on the very brink of extinction. Reports of local people and explorers indicate that *Rhinoceros sondaicus* was quite numerous during the late nineteenth and early twentieth centuries in India, Bhutan, Bangladesh, China, Myanmar, Thailand, Laos, Cambodia, Vietnam, and Malaysia, and on the islands of Sumatra and Java. As an indication of the species' former abundance, Noel Simon and Paul Géroudet quote (but do not further identify) a "Pollack who spent seven years in Assam in the 1860s" (presumably the Fitzwilliam

Thomas Pollock, described by Martin and Martin in 1982). "I never shot the lesser rhinoceros on the right bank of the Brahmaputra and I have no doubt that it still exists," wrote Pollack, "but it is fairly plentiful on the left bank south of Goalparah, where I have killed it. . . . I may here mention that in Assam I shot 44 to my own gun, and probably saw some 60 others slain, and lost and wounded fully as many as I killed."

At 5½ feet at the shoulder, the single-horned Javan rhinoceros cannot easily be mistaken for the Sumatran. The brownish gray skin hangs in heavy folds on the neck, shoulders, and hindquarters, giving the impression that the animal had been assembled piece by piece, jigsaw fashion.

By the 1930s, Udjong Kulon, the preserve at the western end of Java that encompasses some 300 square kilometers (186 square miles), was the "only area where *Rhinoceros sondaicus* still managed to survive," according to Andries Hoogerwerf's 1970 book *Udjong Kulon: The Land of the Last Javan Rhinoceros*. Despite a 1969 Hollywood film entitled *Krakatoa, East of Java*, the volcano that erupted so cataclysmically in 1883 is actually *west* of Java; in fact, what's left of Krakatoa is just off the peninsula of Udjong Kulon. In his otherwise accurate book about the eruption of Krakatoa, Simon Winchester wrote:

> The Krakatoa archipelago is a detached part of Udjong Kulon National Park, most of which includes the peninsula in Java to the south of the islands known as Java Head. The one-horned Javan rhinoceros is plentiful in the park. One of the long-term benefits of the 1883 eruption has been the reluctance of superstitious people to live and settle in large numbers anywhere near the volcano—a reluctance that has protected a very rare species from what would otherwise have been, in the face of population pressure, almost certain extinction.

Although early estimates of the Udjong Kulon rhino population are unreliable, the current numbers hover around twenty, no longer what one would call "plentiful." The declining numbers of *R. sondaicus* led Eric Dinerstein in 2003 to write, "A population of fewer than ten individuals also hangs on in Vietnam. . . . A few Javan rhinoceros may live in the wild, hiding during the day, feeding secretly at night. But for all

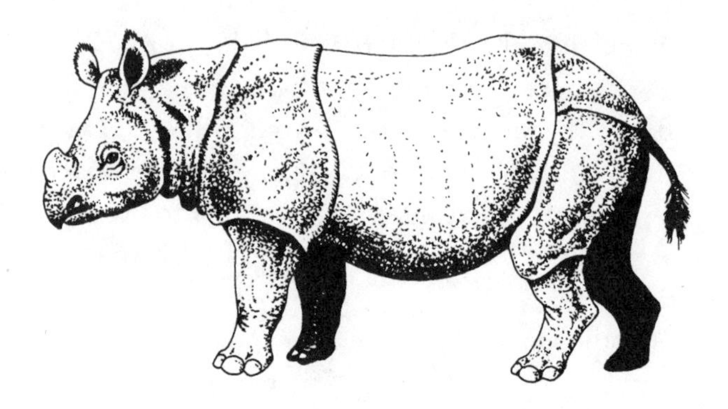

Javan Rhino (*Rhinoceros sondaicus*).

intents and purposes, the species in question are as good as extinct because they no longer play a functional ecological role." Hopes for the survival of one of the world's rarest large mammals were raised by reports that four Javan rhinos have been born since 2001 in Udjung Kulon. The new births were confirmed through an eighteen-month survey carried out by the World Wildlife Fund and the Udjung Kulon national park authority, using camera traps set deep in rhinoceros habitat, DNA analysis of droppings, and tracking to determine the number of animals living in the park.

In Vietnam, "an unpleasant legacy of the prolonged war with the USA is the ready availability of guns and rifles," Charles Santiapillai commented in 1992, and the most serious threat to Javan rhinos is still poaching. During the war, much of the local rhino habitat was contaminated by Agent Orange and other defoliants, evidence of which still exists. The small population of Javan rhinos is still imperiled, but the dense forests might provide some protection.

There were once Javan rhinos also in the Sundarbans, that great swampy, tidal region at the mouth of the Ganges in India and Bangladesh. Swamp deer, wild buffalo, and wild pigs are found there, as are the semi-aquatic Sundarbans tigers. There are many records of Javan rhinos in the Sundarbans, dating from eighteenth- and nineteenth-century hunters, but also from observers who came from Calcutta, the great Indian city that is actually in the Indian Sundarbans, on the Hooghly River. Although the horns of Javan rhinos are usually small, those of the Sundarbans rhinos were virtually nonexistent, making them poor candidates indeed for TCM-related poachers. "They had no trophy worth having, and shooting them was without excuse," said a French sportsman quoted by Kees Rookmaaker (1997b). Horns or no, enough hunters shot the subspecies of the Javan rhino known as *R. sondaicus inermis* in the Sundarbans that they were gone by the beginning of the twentieth century.

Human population pressures in the two protected areas that Javan rhinos are known to exist are now extremely high, with poaching an ever-present threat. A few Javan rhinos live in Cat Tien National Park in Vietnam, just north of Ho Chi Minh City (formerly Saigon). George Schaller and his colleagues in 1990 told of an adult female Javan rhino that had been shot by a tribal hunter in late 1988, northeast of the city. When the hunter brought the horn and hide to town he was arrested and sentenced to a year in jail, which was commuted after two months. The Javan rhinos of Cat Tien, the only ones on the Asian mainland, are considered so important that a study was recently commissioned to assess the status of the remaining rhinos, which are now believed to number only seven or eight. In a 1999 report in *Pachyderm*, Gert Polet and others reported that automatic cameras were set along known rhino routes, resulting in the first-ever pictures of *Rhinoceros sondaicus annamiticus* in the wild. Given the encroachments on habitat and hunting pressure over the past few centuries and the recent increases, it is not surprising that Javan and Sumatran rhinos are endangered; it is more surprising there are any left at all.

The Great Indian Rhino

The Great Asian One-Horned Rhinoceros is especially threatened by the demands of traditional Chinese medicine. Commonly known as the Indian rhino, it is probably the most familiar of all rhinoceroses, with its massive body; short, powerful legs; splayed, three-toed feet; and above all, its "armored" appearance, seemingly comprised of various "plates" and knobs that look more like rivets than any sort of skin formation. In his *Just-So Story* of "How the Rhinoceros Got His Skin," Rudyard Kipling tells of the Parsee who saw that the rhino ("with a horn on his nose, two piggy eyes, and few manners") had eaten his cake, so when the

Indian Rhino (*Rhinoceros unicornis*).

rhino takes off its skin to take a bath, the Parsee fills the skin with cake crumbs, much to the rhino's discomfort:

> Then he wanted to scratch, but that made it worse; and then he lay down on the sands and rolled and rolled and rolled, and every time he rolled the cake crumbs tickled him worse and worse and worse. Then he ran to the palm-tree and rubbed and rubbed and rubbed himself against it. He rubbed so much and so hard that he rubbed his skin into a great fold over his shoulders, and another fold underneath, where the buttons used to be (but he rubbed the buttons off), and he rubbed some more folds over his legs. And it spoiled his temper, but it didn't make the least difference to the cake-crumbs. They were inside his skin and they tickled. So he went home, very angry indeed and horribly scratchy; and from that day to this every rhinoceros has great folds in his skin and a very bad temper, all on account of the cake-crumbs inside.

A big male Indian rhino can be 12 feet long, 6 feet high at the shoulder, and weigh $2\frac{1}{2}$ tons. Both males and females have nasal horns; in Walker's 1991 *Mammals of the World*, the record length for an Indian rhino horn was 20.6 inches. As with all rhinos, the horn of the Indian rhino is a cemented mass of hair that grows on top of the snout. Because it is not actually a part of the skull, the horn can be more easily broken and lost, but it can also be regrown. A rhino's horns are not organs of defense, nor offense, nor for digging. Unlike their African cousins, Indian rhinos normally use their teeth as a major weapon of offense, but they can also charge and trample a person.

By and large, though, Indian rhinos are inoffensive creatures, preferring flight to fight. Though hunted for centuries, there are few records of an Indian rhino attacking a hunter, even when wounded. Their reputed antagonism toward elephants is largely apocryphal; in the wild, at least in Africa, elephants are more likely than rhinos to do the attacking. In captivity, Indian rhinos can become amazingly tame, but male Indian rhinos fight each other for dominance, an inclination some nineteenth-century Indian rulers exploited by staging "rhinoceros fights." One such encounter took place when the Maharajah of Baroda entertained the Prince of Wales in 1875. As described by Rookmaaker, Vigne, and Martin (1998): "[The rhinos] came nose to nose as if to exchange civilities, but the attendants began to excite ill-feelings by poking and patting them alternately, and by horrid yells, and one rhinoceros made a thrust at his friend. . . . They were deluged with cold water to keep up their courage by the attendants, but to no avail, they could not be made to fight each other again. Exeunt two degraded rhinoceroses."

At one time, these behemoths were found throughout the alluvial plains of Pakistan, northern India, Bangladesh, Bhutan, and Nepal, but the population has steadily and steeply declined in the past four hundred years. Throughout the nineteenth and much of the twentieth centuries, Europeans and Asians hunted the great rhinos for sport. In West Bengal and Assam, from 1871 to 1907, the Maharajah of Cooch Behar shot 207, "almost single-handedly sending the Greater One-Horned Rhinoceros to its doom," according to Vivek Menon, a commentary not only on the maharaja's profligacy, but also on how greatly the rhino population had already been reduced. To supply the insatiable thirst for tea in Britain, planters in Assam appropriated vast tracts for tea plantations, eliminating the resident rhinos in the process. Although the Indian government established Kaziranga as a forest reserve in 1908 and officially abolished rhino hunting in 1910, poaching for sale of the horns to TCM practitioners has remained a persistent threat to the rhino population.

What remains is considered endangered and is highly restricted to preserves, but as the price of rhino horn continues to escalate, the number of remaining rhinos—even in protected areas—continues to fall. For people nearby, the temptation is huge. As Geoffrey Ward commented in 1997, "Rhino horn is used in traditional Chinese medicine, and a single horn can bring more than $8,000 on the black market—many times the average income of the people who live around the park."

"To the north of India, for religious, medicinal, and decorative purposes, the Nepalese use more parts of the rhino than any other people in the world, but only a very few rhinos have recently been killed there," E. B. Martin wrote in 1985. Rhino skin was used to make bracelets, earrings, and walking sticks, and in the recent past, more such items, some utilitarian, some decorative, were in vogue. Martin gives this example: "In 1938, when Kiran Shumsher Rana, the son of the Prime Minister then, shot a rhino in southern Nepal, he gave almost all of its skin to a craftsman in Patan to make a spice container, a flowerpot, picture frames, two table lamps, a chandelier, a bowl and a jewel box, all of which he still keeps as very special treasures" (1985a).

The Nepalese believe that drinking rhino urine will relieve asthma attacks, congestion, and stomach disorders. After the horns, hoofs, and hide have been removed from a dead rhino, local villagers converge on the carcass to hack off hunks of meat, which is eagerly brought home, cooked with spices, and eaten to fend off serious diseases. Rhino liver is eaten to cure tuberculosis, and the bones of the kneecap are fashioned into oil lamps to be used in religious ceremonies. Martin again: "Other bones are carved into finger rings to keep away evil spirits. Some are made into charcoal, the fumes of which are believed to cure diseases among penned cattle. In some cases the penis from a dead rhino is taken. I saw a few dried ones for sale in Kathmandu, and I was told that old men boil them in water and eat them to try to cure their impotence." Rhino *penis* may be used for impotence in TCM, but rhino *horn* never is.

By 1968, the rhino population in the Valley of Nepal had fallen to about one hundred animals, and in 1973, the Nepalese government carved out the 600-square-mile Royal Chitwan National Park, about 100 miles southeast of Kathmandu, and posted a special "rhino patrol" to the region. Because the available habitat in Chitwan was found to be

The horn of an Indian Rhino, rescued from poachers in Kaziranga National Park, India.

insufficient for the number of rhinos there, from 1986 to 1999, forty-three rhinos were relocated to the Royal Bardia National Park in the more remote southwest part of Nepal. Crop raiding by rhinos in the adjoining fields led to the occasional rhino-related human casualties in the park, and the foraging animals often faced retaliatory actions by affected farmers. Rhino conservation efforts, including relocation, have generally been a success, however, and as of the turn of the twenty-first century, there were more than five hundred rhinos in Chitwan and fifty-two in Bardia. According to the International Rhino Foundation's Web site (www.rhinos-irf.org) in 2003, "With strict protection from Indian and Nepalese wildlife authorities, Indian rhino numbers have recovered from under 200 earlier in the 20th Century to around 2,400."*

In 1990, the government of Nepal collapsed. Bloody violence ensued, and a state of emergency was declared. It soon became evident that the rhinos in Nepal's national parks were also once again under the gun. From May to November 2000, for example, twenty-nine Indian rhinos were found dead in Chitwan and four in Royal Bardia. To fight the Maoist insurgents, security guards originally detailed to protect the rhinos had been removed from their posts in the national parks, leaving the rhinos at the mercy of poachers who took hides and horns to sell for use in traditional Chinese medicine. By this time, the price for rhino horn had reached 100,000 Nepalese rupees ($1,418) per 100 grams, or 300,000 rupees for a horn that weighed 722 grams (Martin 2001). In mid-2000, however, Gopal Uphadhyay was appointed chief warden, and the main buyer of rhino horn in the Chitwan Valley was arrested. With no outlet for the horn, poaching was virtually halted, and from mid-2000 to February 2001, only one rhino was killed by hunters. Some seven thousand people died from 1997 to 2003 as a result of rebel-government struggles, and the country's tourist-dependent economy was paralyzed. In the summer of 2003, the seven-month cease-fire between

* Eric Dinerstein, now chief scientist and vice president for science of the World Wildlife Fund in Washington, has been described by George Schaller as "the best friend [the rhinos of Chitwan] ever had" for his years of research and rhino conservation efforts. Dinerstein has written a book about the Indian rhino, *The Return of the Unicorns*, which, fortunately, was published while I was working on this one, allowing me to reap the benefits of his scholarship and his dedication to the rhinos.

German seventeenth-century rhinoceros-horn cup, mounted on a silver base with ivory fittings. On one side of the carved horn, an elephant hunt is depicted; on the other, a battle between a rhino and an elephant. The little ivory rhino is a removable stopper. Though this elaborately carved cup could be used for drinking, it was primarily a decorative object.

the Maoists and the government of Nepal was suspended, and hostilities began again. In a Reuters report dated October 7, 2004, we read that "both Nepali government troops and Maoist rebels are abducting, torturing, and killing ordinary people as an eight-year-old conflict escalates across the impoverished kingdom." Bad news for Nepalis; bad news for Maoists—and very bad news for rhinos.

Most of the remaining Indian rhinos not in Nepal can be found in northern India, in the region known as the Terai, at the foot of the Himalayas. In a 1996 TRAFFIC report called "Under Siege: Poaching and Protection of the Greater One-Horned Rhinoceros in India," Vivek Menon tracks the history of poaching since the official ban on rhinoceros hunting was enacted in 1910. There were enough rhinos in India by the 1960s that poachers found it easy to find and kill them: "Between 1968–72 . . . Jaldapara and Gorumara lost 32 rhinoceroses to poaching or a third of the rhinoceroses present in these protected areas; Kaziranga lost 53 rhinoceroses between 1965–69, the first large number lost to poachers, roughly 15% of its population average for the decade." In the 1970s, the figures were lower, but the poachers continued to kill rhinos, especially in protected areas like Jaldapara, Gorumara, and Kaziranga, but the 1980s "marked a resurgence in poaching" and the introduction of two new methods, electrocution and poisoning. From 1980 to 1984, 251 (out of a known total of 1,125) rhinos were killed by poachers, including 61 in the sanctuary of Laokhawa and 125 in Kaziranga. In the next five years, another 232 rhinos were killed for their horns, bringing the total for the 1980s to 483. In the 1990s, the poaching rates accelerated again. From 1990 to 1993, 209 rhinos were killed, almost 15 percent of the remaining population. With the exception of the "black year" of 1983, when the country lost 93 rhinoceroses, never before had more than 60 rhinoceroses been killed in one year, yet in 1992 and 1993, 66 rhinoceroses were poached annually. The most seriously affected reserve of this half-decade was Manas National Park, which officially lost 41 rhinoceroses, 68 percent of its population.

After Assam's Kaziranga and Laokhawa, the rhino population in the West Bengal sanctuaries of Jaldapara and Gorumara is the highest in India. From 1974 to 1989, poaching in West Bengal averaged only one rhino a year, but five rhinos were killed by poachers who lived in the impoverished outskirts of the two sanctuaries. In 1993 the price for rhino horn ranged from $640 to $896 per kilogram. From West Bengal, the contraband horn was often brought to Bhutan, from there to be smuggled into mainland China or Taipei. In 1993, a Bhutanese princess, Dekichoden Wangchuck, the aunt of the king, was arrested at Taipei's Chiang Kai-shek airport, in possession of twenty-two rhino horns. (She had tried to sell the horns in Hong Kong, but when she failed to find a buyer there, she tried Taipei.)

Armed patrols may curtail hunting, but there is little that can be done to counteract the irrepressible forces of nature, such as the devastating floods that ravaged Kaziranga a few years ago. Kaziranga covers some 473 square kilometers (293 square miles) on the floodplain of the Brahmaputra, and in the summer of 1998, heavy rains caused the river to rise, with disastrous results. By September, almost all of the highlands were submerged and most of the antipoaching camps were ruined. Some animals sought refuge on higher dry ground, but many drowned, and others left the park and were killed crossing roads or by villagers and plantation laborers. The drowned carcasses of thirty-nine rhinos were recovered, and another eighteen were found dead of other causes, including one that had been killed by a tiger. It is difficult to determine which is more threatening to the rhinos of Kaziranga, the sporadic but deadly calamities wrought by nature, or the relentless predation of the poachers. The rhinos were nearly obliterated, and six years ago, it was estimated that there were no more than five left in Assam's Manas National Park (Vigne and Martin 1998). Poachers continue to have access to Manas and other parks in Assam, so the best long-term hope for Indian rhinos in India is still Kaziranga, where upward of 1,500 rhinos live. Efforts to combat poaching there have reportedly been vigorous, according to Lucy Vigne and Esmond Martin (1998), and B. K. Talukdar (2002).

Rhino Medicine

"The Indian rhino," wrote Breeden and Wright in their 1996 study of India's wildlife, "is probably the most coveted animal in history. Every part of the anatomy is desired, as a magical cure for a long list of diseases, as an unfailing charm, and in India as an aphrodisiac. To many people in Asia the rhino is a walking apothecary that could cure them of all diseases and keep them so vital they would stay young and potent forever. To the Chinese, the Indian rhino is much better medicine than rhinos from Africa and other parts of Asia." As Alan Rabinowitz wrote in 1995, "It is no small miracle that rhinos still walk the face of the earth. No other group of animals has been so highly prized for so long yet managed to survive human onslaught. The focus of our obsession with this animal has revolved around the protuberance of hardened hair on the animal's head known as rhino horn. Rhino horn played an important role in medieval Chinese medicine, a role that it continues to play in traditional Chinese practices of today."

The use of rhino parts for medicine goes back a long way. In Bernard Read's 1931 translation of Li Shih-chen's 1597 materia medica *Pen Ts'ao Kang Mu*, we read:

> The rhinoceros has a head and belly like a pig. Short feet with 3 toes like an elephant. It is black and the tongue is barbed. There are 3 hairs to each follicle like a pig. There are single-horned, double-horned, and triple-horned species. The *Erh Ya* says the female is like a cow and the male is like a pig. *Kuo P'u* says that the female has only one horn, it is dark colored and weighs 1,000 catties. The male is like a buffalo with the 3 horns in line from the nose, forehead and cranium. . . . The *T'ung T'ien Chüeh* is hollow and the animal can breathe and squirt water through it. . . . The animals were caught by rigging up a rotten wood fence which the animals like to lean against, and falling down as the wood breaks they cannot rise quickly and are easily killed. It is said to shed its horn each year, and bury it in the ground. If carefully replaced by wooden imitations 3 times the animal will continue to plant its horn in the same place year by year.

In his 1974 *Introduction to the Mammals of Singapore and Malaya*, John Harrison repeats these qualities of rhino horn but says, "Such accounts, which appear in Chinese herbals, are clearly muddled hearsay, but the material of the horn was, and is, obviously used." Relying largely on Read's translation of the materia medica, Harrison commented, "Gossip says that it is a powerful aphrodisiac, but the uses listed are as an antidote to poison, to cure possession by devils; to keep away evil spirits and miasmas; to remove hallucinations, bewitching nightmares, infantile convulsions and dysentery. . . . All this for a few hundred dollars an ounce! No wonder the market is flooded with imitation rhino horn."

Even though many Westerners believe that rhino horn is used as an aphrodisiac in certain Asian countries, it is in fact used to cure almost everything *but* impotence and sexual inadequacy. In Read's translation of *Pen Ts'ao Kang Mu*, the complete section on rhinoceros horn ("the best is from a freshly killed male animal") reads as follows, with no mention of any aphrodisiac qualities:

> It should not be taken by pregnant women; it will kill the foetus. As an antidote to poisons (in Europe it was said to fall to pieces if poison were poured into it). To cure devil possession and keep away all evil spirits and miasmas. For gelsemium [jasmine] and snake poisoning. To remove hallucinations and bewitching nightmares. Continuous administration lightens the body and makes one very robust. For typhoid, headache, and feverish colds. For carbuncles and boils full of pus. For intermittent fevers with delirium. To expel fear and anxiety, to calm the liver and clear the vision. It is a sedative to the viscera, a tonic, antipyretic. It dissolves phlegm. It is an antidote to the evil miasma of hill streams. For infantile convulsions and dysentery. Ashed and taken with water to treat violent vomiting, food poisoning, and overdosage of poisonous drugs. For arthritis, melancholia, loss of the voice. Ground up into a paste with water it is given for hematemesis [throat hemorrhage], epistaxis [nosebleeds], rectal bleeding, heavy smallpox, etc.

"People from northern India, Burma, and northern Thailand," wrote Martin (1987a), "consume rhino blood, urine, and penises as aphrodisiacs. Rhinos are capable of copulating for as long as 90 minutes, with the male having orgasms at two-minute intervals. Perhaps this 'feat' is the basis for belief in the aphrodisiac powers of the rhino's penis." Because it was believed to provide such a pharmacological bounty, it is perhaps superfluous for rhino horn also to serve as a love potion. How then did rhino horn acquire its aphrodisiacal reputation? Probably from Western writers who had only a passing acquaintance with Chinese traditional medicine, such as the great white hunter with the eponymous name, who, in 1952, wrote:

> The horns are worth thirty shillings a pound or more—ten shillings more than the finest grade of ivory. These horns are used for a curious purpose. Orientals consider them a powerful aphrodisiac and there is an unlimited demand for them in India and Arabia. No doubt any man who has a harem of thirty or more beautiful women occasionally feels the need for a little artificial stimulant.

In his 1962 study of the animals of East Africa, C. A. Spinage seemed to share the belief that Asians were interested in the horn as an aphrodisiac and were willing to pay handsomely for it: "On account of mysterious aphrodisiac properties attributed to the horn by certain Asiatic peoples, the Rhino has been sorely persecuted. . . . With its horn fetching the present high price the prospects of its continued survival in the face of the poachers' onslaught are not very bright."

The anthropologist Louis Leakey also shared this misunderstanding. In his 1969 book on African wildlife, he commented that rhinos were "in grave danger from poachers because rhino horn commands a high price in the Far East, where it is rated as an aphrodisiac." Similarly, in their 1970 book *Last Survivors*, Noel Simon and Paul Géroudet wrote, "It is impossible to shake the firmly held belief, prevalent throughout the East, in the infallibility of rhino horn as a powerful aphrodisiac, and most Asians will go to any length and pay almost any price to obtain it." And in *S.O.S. Rhino*, C. A. W. Guggisberg asserted that "the superstition that has done more harm to the rhinoceros family than all others is undoubtedly the Chinese belief in the powerful aphrodisiac properties of the horns. Through the centuries untold generations of aged gentlemen have been imbibing powdered rhino horn in some appropriate drink, hoping to feel like a twenty-year-old when next entering the harem!"*

Of this sample of writers, Hunter was hardly a conservationist, but Leakey, Spinage, Guggisberg, and Simon and Géroudet believed in protecting animals, so if they were responsible for spreading the myth of rhino horn as a sexual amplifier, it would countermand everything they stood for. Nevertheless, wrote Martin and Martin in *Run Rhino Run*, "it seems that the myth from East Africa moved to Asia via the conservationists. Having heard or read that that the aphrodisiac value ascribed to rhino horn was the reason why the Chinese were buying it in vast quantities from East Africa, they offered the same explanation for the poaching of Asiatic rhinos, which were even more in demand on the Chinese market."

Make no mistake: those people who use rhino horn to cure medical ailments really believe it works. That's what drives up the demand on which the poachers thrive. As Ann and Steve Toon commented in 2002, "For practitioners of traditional Asian medicine, rhino horn is not perceived as a frivolous love potion, but as an irreplaceable pharmaceutical necessity." And Eric Dinerstein (2003), concurs: "In fact, traditional Chinese medicine never has used rhinoceros horn as an aphrodisiac: this is a myth of the Western media and in some parts of Asia is viewed as a kind of anti-Chinese hysteria."

In the 1960s and 1970s, Hong Kong was the world's largest importer of rhino horn. Although the government officially banned all imports in 1979, rhino horn was smuggled in from Macao, Burma, Indonesia, Malaysia, India, Taiwan, and South Africa. At the 1987 CITES meeting in Ottawa, participating parties agreed to abate the rhino crisis by closing down the trade in rhino products completely. British Prime Minister

* "When we returned to camp," wrote Hunter, "I tried some of the horn myself to test its powers. I closely followed the recipe given me by an Indian trader: take about one square inch of rhino horn, file it into a powder form, put it in a muslin bag like a tea bag, and boil it in a cup of water until the water turns dark brown. I took several doses of the concoction but regret to report that I felt no effects. Possibly I lacked faith. It is also possible that a man in the bush, surrounded by nothing by rhinos and native scouts, does not receive the proper inspiration to make the dose effective."

Margaret Thatcher promised the ban would take effect later that year. This never happened in an effective way, of course, but there were suggestions that substitutes for actual rhino parts might suffice for TCM. Scientists at the China Pharmacological Institute proposed using buffalo horn (made of keratin, as are rhino horns), for example, and the general manager of China National Health Medicines Products told Esmond Martin that all their new medicines now used buffalo horn instead of rhino horn. "Regrettably," he wrote, "not all the factories in China have switched to water buffalo horn . . . and they were still utilizing old stocks of rhino horn" (Martin 1989a). In the section on "Heat-clearing, blood-cooling medicinals" in Wiseman and Ellis's 1996 *Fundamentals of Traditional Chinese Medicine*, we find the admission that all those rhinos didn't have to be killed at all. After a list of all the symptoms that rhinoceros horn (*xi jiao*) can alleviate, there is this note: "The rhinoceros is an endangered species. Please use water buffalo horn as a substitute."

Chryssee and Esmond Martin observed in a 1991 book that "Taiwanese self-made millionaires are notorious for their conspicuous consumption of rare and exotic wildlife, and the Chinese traditional adage that animals exist primarily for exploitation is nowhere more pronounced than on this island." During a visit to Taiwan three years earlier, they found that most of the rhino horn for sale there had come from South Africa. Two years later, they found that, despite a ban on imports, domestic sales of smuggled horn were still high. The demand for Asian horn in particular was "increasing and wealthy Taiwanese, aware that prices will rise even higher as rhinoceros numbers decline, are buying it as an investment." In those regions where rhino-horn products are dispensed—legally or illegally—the most popular medicines are used for tranquilizers, relieving dizziness, building energy, nourishing the blood, curing laryngitis, or simply, as the old snake-oil salesmen would have it, "curing whatever ails you."

In Bensky and Gamble's 1993 *Chinese Herbal Medicine*, we read that cow bile may be substituted for bear bile, "because of the endangered status of many bear species," but rhino horn (*cornu rhinoceri*) can be obtained from any species of rhinoceros, "most commonly the Asian species *Rhinoceros unicornis*, *R. sondaicus*, or *R. sumatrensis* are used;

sometimes the African species *R. simus* and *R. bicornis* are used." To such Chinese herbal medicine specialists, the rarity of some species of rhinoceros may be unimportant (and the misidentification of their names of no consequence), and while some practitioners might allow a substitute for bear bile, nothing, it appears, can take the place of *cornu rhinoceri*.

The Rhino-Horn Trade

Since the late 1970s, Esmond Bradley Martin has taken the lead in research on the rhino-horn trade. Born in 1941 in New York City and educated at the universities of Arizona and Liverpool, he went to Africa in 1971 to study the dhow trade for a doctorate in geography. Even before his numerous papers on rhinos appeared, he and his wife, Chryssee, published the influential *Run Rhino Run* in 1982, with an introduction by Elspeth Huxley.

"It was largely because of Bradley Martin," wrote Raymond Bonner in 1993, "whose shock of unruly and prematurely white hair makes him instantly noticeable in any crowd, that the rhino had been declared an endangered species under CITES, and international trading in rhino horn became illegal in 1973." Now based in Nairobi, Martin has made a career out of rhino research; he has seen living rhinos wherever they can be seen and has visited medicine factories, shops, and pharmacies for firsthand observations in such places as Mombasa, Yemen, Taipei, Hong Kong, Macao, Singapore, Kuala Lumpur, Bangkok, Borneo, Japan, Beijing, and various other cities in China. Indeed, *Run Rhino Run* would have been an ideal primer for the study of rhinos in today's world—if only things hadn't got a lot worse since its 1982 publication. Through Martin's voluminous publications—often in *Pachyderm* but in other international journals as well—one can follow the trails of crass commercialism, international intrigue, pharmacological preparations, and the valiant efforts to rescue rhinos by various conservationists, one of the most important of whom is Martin himself. In the early 1990s, Martin was named the United Nations Special Envoy for Rhino Conservation, and he has worked with (and his efforts have been supported by)

Kes Hillman Smith and Esmond Bradley Martin measure rhino horns collected by officials of South Africa's Pilanesberg National Park in 1981. Soon after this picture was taken, these horns were stolen and probably ended up in East Asia.

virtually every conservation organization in the world, from the WWF and the IUCN to the Wildlife Conservation Society and the National Geographic Society.

Even before the recent increase in demand for rhino horn for TCM, rhinos were slaughtered en masse in Africa for their horns, which had backed them toward the precipice of extinction in the wild. For *Run Rhino Run*, the Martins researched the history of what they called "The Rhino Business" and uncovered some astonishing statistics. Seyyid Said, the sultan of Muscat, moved his capital to Zanzibar in 1840, and he and his successors "presided over a massive slaughter of rhinos never equaled before or since. . . . One of the foremost Zanzibar historians has estimated that the towns of Mafia and Bagamoyo used to receive from the interior of East Africa between 5,500 and 8,000 kilos [5.5 to 8.8 tons] of rhino horn per year in the 1840s." All told, the Martins estimated that "between 1849 and 1895, an average of 11 tons of rhino horn a year was exported from East Africa, which meant that 170,000 rhinos must have been killed to supply the trade during this 49-year period." Most of these horns went to India to be carved, or to China to be incorporated into medicines, but a startling 30 percent in the early twentieth century went to Britain. What in the world did the Brits do with all that rhino horn? They used it, it turns out, for everything from decorative carvings to the tops of riding crops, walking sticks, door handles, and interior fittings in carriages. "Polished rhino horn looks almost like amber and is not difficult to carve," wrote the Martins. "It had a prestige value too: since rhinos were still regarded by most Europeans as exotic beasts, a dandy with something made from rhino horn could almost certainly count on attracting the attention of his peers."

The scarcity of rhinos today, and the corresponding intermittent availability of rhino horn, only drives the price higher, and intensifies the pressure on the declining rhino populations. For people whose annual income is often far below the subsistence level, the opportunity to change one's life by killing a large, ungainly, and otherwise seemingly "useless" animal must be overwhelming. How much is rhino horn worth? In Nowak's 1991 revision of *Walker's Mammals of the World*, we read: "*R. unicornis* is jeopardized by loss of habitat to the expanding human population and illegal killing, especially in response to the astonishing rise in the value of the horn. The wholesale value of Asian rhino horn increased from US $35 per kg [2.2 pounds] in 1972 to $9,000 per kilogram in the mid-1980s. The retail price, after the horn has been shaved or powdered for sale, has at times in certain East Asian markets reached $20,000–$30,000 per kilo. In contrast, in May 1990, pure gold was worth about $13,000 per kilo."

Estimates of price for rhino horn used to make Yemeni daggers is controversial. A letter from Esmond Martin in the July 2004 issue of *National Geographic* took issue with a previously published article that cited $60,000 as the price for a kilogram of black rhino horn used to make Yemeni daggers. The going price was actually far lower, $1,200, Martin said. By "stating this very high price for rhino horn," he went on, "poachers and traders will have a great economic incentive to kill rhinos and send the horns to Yemen." Even if the lower figure is correct, $1,200 is a huge amount of money by third-world standards. In his many publications on the rhino horn trade, Esmond Martin himself has quoted different prices at different times, and in different countries. For example, in Yemen in 1987, he found that traders would pay from $800 to $1,000 per kilogram for rhino horn. In a 1989 paper about the rhino trade in Borneo, Martin tells of a poaching gang in Sarawak—the rhinos there are Sumatran—that received $7,300 for a pair of horns, nails, and a good portion of hide and a few bones. In Taiwan in 1989, customers were willing to spend $40,000 per kilogram, "the highest retail price in the world," and in a 1990 study of rhino-horn consumption in South Korea, Martin wrote that the price in 1980 was $1,436 per kilogram; and by 1988, it had risen to $4,410.

As living rhinos become increasingly unavailable in the wild, those who would market horn products must turn to other sources. In April 1990, Martin visited Beijing's largest pharmaceutical factory, Tong Ren Tang, where he found more than 10 tons of rhino horn stockpiled for future use. When he was shown the storerooms—the Chinese do not think that what they are doing is wrong and are not reluctant to have Westerners see their operations—he also saw "plastic bags, crates and boxes containing chips, powder, whole horns, together with the most amazing form of stock to be used for making medicine, that of antique rhino horn carvings . . . antique plates, cups, libation bowls, brush holders and figurines. [He] even saw quite a few Sumatran, Indian, and Javan carved horns" (Martin 1990).

Keratin—the major protein components of hair, wool, nails, horn, hoofs, and the quills of feathers—in rhinoceros horn is chemically complex and contains large quantities of sulfur-containing amino acids, particularly cysteine, but also tyrosine, histidine, lysine, and arginine, and the salts calcium carbonate and calcium phosphate. Rhino horns are composed primarily of keratin, but so too are rhino nails. Three to a foot, for a grand total of twelve per rhino, the nails can also be shaved or powdered for pharmaceuticals. You cannot carve a jambiya handle from a toenail, but shaved or powdered rhinoceros keratin, with all its believed powers, might be beneficial regardless of which part of the rhino it comes from. In a Bangkok medicine shop, Lucy Vigne and Esmond Martin (1991) found that "a large collection of rhino products, usually from the Sumatran rhino . . . occasionally nails and even dried penises are for sale." A year later, in a study of the Bangkok market for rhino products, Martin noticed a dramatic increase in the price of Sumatran rhino nails, from $1,487 per kilogram in 1986 to $13,905 per kilogram in 1992. The horn, however, remains the ne plus ultra of rhino products; its favored status built of long tradition is attributable in part to a strong historical continuum with the original myth of the unicorn.

Unicorn or rhino, the horn is used in TCM because people believe it works. Westerners have largely rejected the TCM notion that it can reduce fevers, but it has been demonstrated that in very large quantities it can indeed reduce fever in rats (But et al. 1990), and the rats obviously cannot be influenced by the placebo effect. As long as people believe in the medicinal properties of rhino horn they will continue to use it. That the medicines may not cure or ameliorate the conditions for which they are prescribed in TCM cannot be accurately assessed by a Western-only perspective, say many TCM adherents, so an argument on grounds of efficacy will likely fall on deaf ears. But the argument that killing rhinos for their horns or toenails endangers the species is a powerful one, and it is probably the only one that has a chance of saving the remaining rhinos.

Tibetan medicine, somewhat different from Chinese, also employs rhino horn, even though the Dalai Lama has expressly condemned the illegal killing of animals. Vivek Menon's comprehensive 1996 report (*Under Siege: Poaching and Protection of Greater One-Horned Rhinoceros in India*) concludes that "the poached rhinoceros horn is used to a small extent in the domestic market in India, for Tibetan medicinal and other uses, but that the main demand is medicinal use in the markets of East Asia and Southeast Asia." Throughout those markets, the trade in rhino horn for medicinal purposes is a very big business, but because much of it is conducted through various black markets, its true magnitude may never be known.

In a 1989 article in *Pachyderm*, Lucy Vigne and Esmond Martin suggested that "Taiwan may have become the world's largest entrepot for

It is not immediately evident what this object was made to contain, but whatever its function, its cost was one rhino foot. The bronze rhino finial was sculpted by James L. Clark of the American Museum of Natural History.

African and Asian rhino horn." Lacking United Nations recognition, Taiwan cannot sign CITES. According to a 1992 TRAFFIC report by Kristin Nowell, Chyi Wei-Lin, and Pei Chia-Jai, "7,281 kg (16,000 pounds, or 8 tons) of rhino horn were legally imported between January 1972 and August 1985. It is likely that over the same period additional quantities were brought in undeclared in order to avoid customs duties." Most of it is consumed—in very small quantities—to reduce fever, because many among the 20 million Taiwanese prefer the remedies offered by Chinese traditional medicine to those of modern medicines with their uncertain side effects, or choose to use both types. Although medicines made from Asian rhinos are preferred by Taiwanese, pharmacies are also stockpiling the cheaper African horn as a hedge against the time when the Asian rhinos become too rare to harvest. But, Menon wrote, "Asian rhinoceros horn is far more porous and soft than African horn. . . . A number of dealers . . . claim that it is almost impossible to carve an Indian rhinoceros horn (and it would, in any case, be much more likely to be sold for Oriental medicine), whereas an African one may be carved with relative ease. Conversely, it is easier to powder Indian horn than African horn."

The Taiwanese make up much of the market for horn imported to Asia from South Africa, Mozambique, Tanzania, and Zimbabwe—wherever black rhinos can still be found. Threats of jail time or stringent fines apparently do not have much of an effect on rhino poaching in Africa; in a 1989 article in *Pachyderm*, conservationist David Western wrote, "The unabated decline in African rhinos during the 1980s shows that poaching has defied all efforts to ban the horn trade. . . . The prospects are likely to worsen as the task of detecting fewer and fewer horns entering the market becomes more formidable and price incentives rise."

Like the Taiwanese, many Koreans are devoted practitioners of traditional medical arts and are prepared to import substantial amounts of substances not naturally found in their country. Korean traditional medicine is based on the Chinese version, which is said to have come to Korea during the sixth century. "Rhinoceros horn," wrote Judy Mills in 1993, "is an ingredient in five . . . medicines still popular among doctors of Oriental medicine in Korea today. These rhinoceros horn derivatives are used to treat maladies including stroke, nosebleeds, dermatitis, headache, facial paralysis, high blood pressure, and coma. The most popular of these medicines is *Woo Hwang Chang Shim Won*, a medicine ball made from rhinoceros horn, musk, cow gallstones, and a number of herbs." In 1992, the U.S. government threatened to impose sanctions via the Pelly Amendment* on South Korea for failure to police the trade in rhino horn. In response to the threat—the Pelly Amendment was not actually employed—the price of rhino horn in South Korea doubled. Among the some seven thousand doctors licensed to practice Korean medicine in South Korea (no figures are available for North Korea), there was little diminution of prescriptions written for *Woo Hwang Chang Shim Won* after 1992. In fact, it is not clear that the use of rhino horn for medicinal purposes has decreased at all.

Rhino horn has been an integral component of TCM for thousands of years. It matters little where the rhinos come from; the horn of a rhinoceros from any continent may be used for medical purposes. In East Africa—primarily Kenya, Uganda, and Tanganyika (now Tanzania)—statistics on rhino-horn harvesting have been kept since 1926. Over this period, most of the rhinos killed were black rhinos, although the "harvesters" would not pass up a white rhino if it appeared in their gunsights. During the 1930s, according to Nigel Leader-Williams (1992), declared exports from East Africa (then under British rule) averaged about 1,600 kilograms (3,520 pounds) per year, which meant the death of some 555 black rhinos annually. During World War II, the numbers soared to 2,500 kilograms (5,500 pounds), for which approximately 860 rhinos died each year. During the 1950s and 1960s, the auction houses reported about 1,800 kilograms (3,960 pounds) per year, which would have entailed the death of about 600 rhinos every year in that period. In the 1970s, the numbers skyrocketed again, to 3,400 kilograms (7,480

* The Pelly Amendment—officially the Pelly Amendment to the Fishermen's Protective Act of 1967—is a restriction on importation of fishery or wildlife products from countries that violate international fishery or endangered or threatened species programs. It has been used in fisheries and in whaling disputes, but because it can be applied to any "threatened species programs," it can be used against any country that the United States determines is in violation of international treaties such as CITES.

pounds), and every year in that decade, 1,180 rhinos died. Leader-Williams identifies the Far East's primary consuming nations as Hong Kong (which was separate from the People's Republic of China until 1997), mainland China, Taiwan, Singapore, Japan, South Korea, Peninsular Malaysia, Sabah Malaysia, Brunei, Macau, and Thailand, while the major Asian importers of African rhino horn were, not surprisingly, the first three on this list—mainland China, Hong Kong, and Taiwan.

More or less at the same time that Judy Mills was compiling her report on the rhino-horn trade in South Korea, Leader-Williams (now director of the Durrell Institute for Conservation and Ecology at the University of Kent) was assembling a review of the world trade in rhino horn and determining successes in attempts to halt it. Curiously, he buys into the fallacious notion that rhino horn is (or was) common as an aphrodisiac, for he writes, "The pharmacological efficacy of rhino horn as an aphrodisiac can only be guessed at. However, its psychological value may well be all important and has some basis both in the shape of rhino horns and in the long courtship and staying power of copulating rhinoceroses, which take up to an hour from intromission to ejaculation" (1992). Even without aphrodisiacal properties, however, rhino horn is one of the mainstays of TCM, and its collection has been responsible for the death of tens of thousands of rhinos around the world.

Dehorning the Rhino

When African rhinos were considered big-game animals or an impediment to settlement, the slaughter was enormous. Indeed, so many rhinos were killed during the colonial era that protecting the small percentage that survives has assumed a much greater importance. Documenting the poaching and illegal trade in Namibia's Etosha National Park from 1980 to 1990, Esmond Martin learned that poachers had shot fifteen animals in the park in 1984, another seven in 1987, and twenty-three in 1989. Another seven rhinos were poached in 1989 in northwestern Namibia's Kaokoveld district, for a total during the ten-year period of what was estimated as at minimum sixty-four black rhinos and a few white rhinos. In his 1994 paper on rhino poaching in Namibia, he wrote:

> About five small gangs, usually consisting of only two people, spent between one and three days in the park. They shot the animals during the day; and as well as the horns, for the first time in recent years, they also took some skin. The contact men hoped to sell a pair of horns to Portuguese and Angolans in Windhoek for 2,000 to 4,000 rands ($760 to $1,520.)

If there is no effective way to limit or control the illegal Asian trade in rhino horn for TCM, perhaps there is a way to keep the poachers from killing the rhinos in the first place. What if they had no horns? The idea of removing the horns of rhinos so that poachers would have no cause to kill them has been circulating for some time. In 1989 Leader-Williams noted that

> dehorning has been discussed as a measure to prevent poaching since the 1950s, [but] until now it has been discarded in most areas of Africa for several reasons. First, the cost of dehorning several thousand rhinos over tens of thousands of square kilometers would be extremely expensive. Second, the two African species, the black and the white rhino, use their horns in sparring and to defend calves against predators such as lions and spotted hyenas. Hence, hornless rhinos may be unable to maintain their social status or to rear their calves successfully. As important, most black rhinos live in thick bush, and a poacher sighting only a part silhouette could shoot before finding out his quarry is hornless.

Despite such warnings, in 1989, conservation officials in Namibia's Damaraland dehorned twelve rhinos to see whether it would reduce poaching. This region was chosen as the site of the experiment because it has none of the rhino's natural predators, and there were sufficient tribesmen to act as game guards. In *Run Rhino Run*, Martin and Martin described the process as a hare-brained scheme that would be "fantastically expensive and would also have to be extremely thorough—and repeated at intervals since rhino horns do grow back," but in 1994, Esmond Martin wrote that in fact "the dehorning exercise has been successful . . . two attempts were made to kill rhinos, but once [the poachers] saw that the rhinos had no horns, the poachers left them alone." And Leader-Williams suggested now that a program of cauterizing the horn could be tried, to prevent regrowth of the horns.

Poachers and other "harvesters" of rhino horn usually kill the rhino and saw off the horn. But if the horn can be removed without otherwise harming the rhino, doesn't that suggest harvesting the horns and not the rhinos? There are no nerves in rhino horn, so there is no pain necessarily affiliated with its removal, but of course the animal has to be forcibly immobilized or shot with a tranquilizing dart before the horn can be sawn off. From the rhino's point of view, the stress involved in dehorning might be preferable to death. Along with dehorning, the question of a legalized trade in rhino horn, where the horns would be collected and then the animals released, was also suggested.

In an attempt to find out if this controversial process was practicable, biologists Joel Berger and Carol Cunningham went to Namibia to see for themselves if dehorning would actually protect rhinos from poachers. They also wanted to learn if the removal of a male rhino's horn would affect its breeding potential: would a hornless male become subordinate to horned males and thus be less available for breeding? Over three field seasons from 1991 to 1993, Berger and Cunningham studied the Namib rhinos, often dodging lions, elephants, and even the rhinos themselves. They had wondered if the dehorning process would backfire because hornless mother rhinos would be unable to defend their calves against lions or hyenas. They noted "that dehorned animals have thwarted the advance of dangerous predators"(1994b). They also wrote that "on the basis of over 200 witnessed interactions between horned rhinos and spotted hyenas and lions, we saw no cases of predation." But Berger and Cunningham concluded that hornless mothers—or those with regenerating horns—were less capable of protecting their offspring than those with horns. Perhaps more useful than dehorning, they wrote, would be the isolation of breeding populations of black rhinos far from poachers, which would allow the decimated rhino populations to recover, rather like the efforts that resulted in the rescue of two otherwise critically endangered species, Père David's deer and Przewalski's horse.*

In a 1993 paper with Malan Lindeque, the director of Etosha's Ecological Institute, and A. A. Gasuweb, Berger and Cunningham noted that "horn regrowth is rapid, averaging nearly 9 cm of total horn per year, and because poachers fail to discriminate between large- and small-horned rhinos, recently dehorned animals might not be immune from poaching." So dehorning didn't necessarily protect rhinos from poachers, but it didn't do them any harm either. According to a 1991 *National Geographic* article by Des and Jen Bartlett, "At least 250 rhinos roamed Kaokoveld [Namibia] until the early 1970s. By then poachers, primed by world demand from people who still believe in rhino horn as a medicinal panacea, had gone on a rampage that left only about 65 rhinos standing. . . . To help save their lives a controversial technique renders them valueless to poachers: In two high risk areas rhinos have been tranquilized and their prized horns cut off and stored away." In a review of Cunningham and Berger's book, *Horn of Darkness*, Brian Bertram wrote, "The answer to the question 'Does rhino dehorning work?' is a gloomy one. So far, nothing has been shown to work that will prevent these magnificent animals from being needlessly slaughtered. . . . Cunningham and Berger can at least show that they tried, and did so valiantly."

If hunting was once the main cause of rhino population decline, in recent times, poaching rhinos for their horns has become the primary cause of the worldwide decline. But what if there were a *legal* market for rhino horn? Michael 't Sas-Rolfes of South Africa (1997) believes that

> it is wrong to assume that establishing a legal market is risky. It may in fact be riskier to leave the rhino horn trade solely in the hands of illegal operators. Establishing a legal market could provide a further advantage: a substantial source of revenue for conservation

* Père David's deer, originally an inhabitant of China, was saved from extinction by a breeding program in Europe, led by the Duke of Bedford; a thriving population developed, and recently a small group has been reintroduced to a reserve in China (Hu and Jiang 2002). Przewalski's horse, found in the Mongolian steppes and deserts, was saved from extinction by a captive breeding program (also led by the Duke of Bedford) and, like the Chinese deer, has now been reintroduced in its original habitat.

> agencies. Even more so than with ivory, the potential to fund field protection with the proceeds from legal rhino horn sales is considerable, and could be of great benefit to conservation generally. Conversely, if poaching pressure increases, and conservation budgets continue to shrink, the outlook for rhino protection is bleak.

"Legal trade" in rhino horn may be an oxymoron. It would function as a conservation measure only if it was rigorously controlled, with a clear chain of custody and full control at all points, from source to final sale. But if the Asian wildlife trade, rife with corruption at every level, is any example, no amount of government intervention can circumvent the black market that has been shown to control every aspect of supply and demand. As long as a single rhino horn is worth as much as two years' normal wages, poachers will continue to operate outside the law.

In their 1997 book, Cunningham and Berger noted that as of 1992, Zimbabwe had the largest surviving numbers of black rhinos in the world—approximately 2,000 animals. Within a year, the number had fallen to 300–500, and no one knew why. Poaching was the most credible possibility, but others existed, including drought and starvation. Black rhinos were under siege, horned and dehorned alike. To forestall the extinction of the black rhinos of Zimbabwe, the government decreed that all rhinos in that country would be dehorned and the horns stored for safekeeping. But within a year, the program was abandoned when it was seen that dehorning did not deter poachers, and Zimbabwe's remaining rhinos were moved to smaller areas known as Intensive Protection Zones.

The current government of Zimbabwe might not be the best custodian for rhinos or rhino horns. In an October 25, 2003, article in the *New York Times*, Michael Wines wrote:

> Hunting and tourism once pumped millions of dollars into Zimbabwe's economy each year, sustaining wildlife management programs on millions of acres of private scrubland too arid or rocky for commercial farming, but ideal for photographic safaris and big-game hunts. Zimbabwe's decision to confiscate most of that land from its white owners, and then redistribute it to peasants and political supporters has had an unexpected result: thousands of hungry families

> on land too poor to support crops have turned to poaching as their prime source of food and income. . . . Precise figures do not exist. But by estimates from several conservationists, former landowners and opposition politicians, as many as two-thirds of the animals on Zimbabwe's game farms and wildlife conservancies have been wiped out.

Whether they were horned or dehorned, nothing seemed to work to protect Africa's rhinos from poachers. But decades earlier, when somebody learned that there was an animal, not endangered, whose horns might be substituted for those of rhinos in TCM, it appeared that the rhinos might be offered a reprieve.

Can the Saiga Save the Rhino?

Around the late 1980s, when it was recognized that rhinos were being slaughtered out of existence, conservationist groups like the WWF actively encouraged the hunting of the saiga, promoting its horn as an alternative in traditional Chinese medicine to the horn of the endangered rhino. The results were disastrous.

What is a saiga? It is *Saiga tatarica*, a funny-looking antelope of the Russian and Mongolian steppes, with a bulging proboscis like that of a tapir, a chunky body, spindly legs, and a yellowish gray coat that turns lighter in winter. Only the males bear horns, and these are 8 to 10 inches long, semitranslucent, flesh-colored, and ringed on the lower two-thirds. The saiga's exaggerated nasal passages are an adaptation to the swirling dust of its arid habitat; the nostrils point downward and the passages are lined with a complex arrangement of hair, glands, and mucous membranes that filter, warm, and moisten the air as it is inhaled. Saigas wander for miles every day, marching with their heads low to the ground, their specialized noses filtering out the stirred-up dust. At the end of April, the males start their seasonal spring migration, forming herds of up to two thousand animals and setting out northward. In the meantime, females wander in huge congregations to a suitable birthing ground where they drop all their calves within about a week of one

Saiga (*Saiga tatarica*).

another. Eight to ten days after the calves are dropped, the females and new babies set off after the males, in groups that may exceed one hundred thousand animals. Once the migration is finished, the saigas disperse in smaller herds, only to recongregate in the autumn and move en masse back south. A timid species, the saiga can be easily startled, precipitating a panicked stampede, even in the huge migrating herds. Saigas flock together even when being shot at, so they are particularly easy to kill in large numbers.

Hunters on the steppes killed them for meat and furs, and by the time of the Soviet revolution, there were only a few thousand saigas left. To forestall their total eradication, the Soviets protected saigas in Europe in 1919 and in Soviet Central Asia in 1924 by banning individual hunting and commercial harvests, but in the 1950s, commercial harvests by local groups began again. Saigas were shot from vehicles, which spooked them and worked poorly for large-scale harvesting. In some cases, they were "jacklighted" at night; dazed by bright spotlights and shot with buckshot. Despite such practices, the antelopes made a remarkable comeback, and by 1957 a loose herd estimated to number between 150,000 and 200,000 animals was observed east of the Caspian Sea. Although the saiga population on the steppes in 1993 had risen to over a million, Fred Pearce noted in a *New Scientist* article that fewer than 30,000 were left in 2003, most of those females (2003a). The males were being killed for their horns, which can fetch as much as $100 per kilogram (three pairs of horns weigh about a kilogram), helping to fill the almost insatiable demand among Chinese who believe the horns have beneficial pharmacological properties.

There are four populations of saiga in Kazakhstan: Kalmykia, Ural, Ustiurt, and Betpak-Dala; a small population also exists in western Mongolia. The Kazakhstan populations were not helped by a biological weapons experiment—probably at Stepnogorsk, the most notorious of the Soviet Union's germ warfare testing facilities—where thirty thousand saigas were killed when the wind shifted unexpectedly (Miller, Engelberg, and Broad 2001). All saiga populations are declining, but in Betpak-Dala the numbers plummeted from half a million in 1993 to four thousand in 2000, a drop of 99 percent. Between 1993 and 1998, the million-plus population was essentially halved as the horn-bearing males were hunted, and, as Eleanor Milner-Gulland and her colleagues concluded in 2001, the lack of males in Kalmykia is causing dramatically reduced conception rates, which, in addition to the high hunting mortality, could lead to population collapse.

Ironically, rhino advocate Esmond Bradley Martin had spearheaded the movement to substitute saiga horn for rhino horn. But when he realized that the saiga was close behind the rhino on the fast track to extinction, he publicly recanted. As Milner-Gulland et al. wrote in *Nature* in 2003: "Horns are borne by males and are highly favored in traditional Chinese medicine, which has led to heavy poaching since the demise of the Soviet Union, when the Chinese-Soviet border was opened." It has now been estimated that less than 5 percent of the population survives, and that may have taken the saiga beyond the point of no return. As of 2002, with the wild Bactrian camel and the Iberian lynx, the saiga has been classified as "critically endangered" by the World Conservation Union (formerly the IUCN).

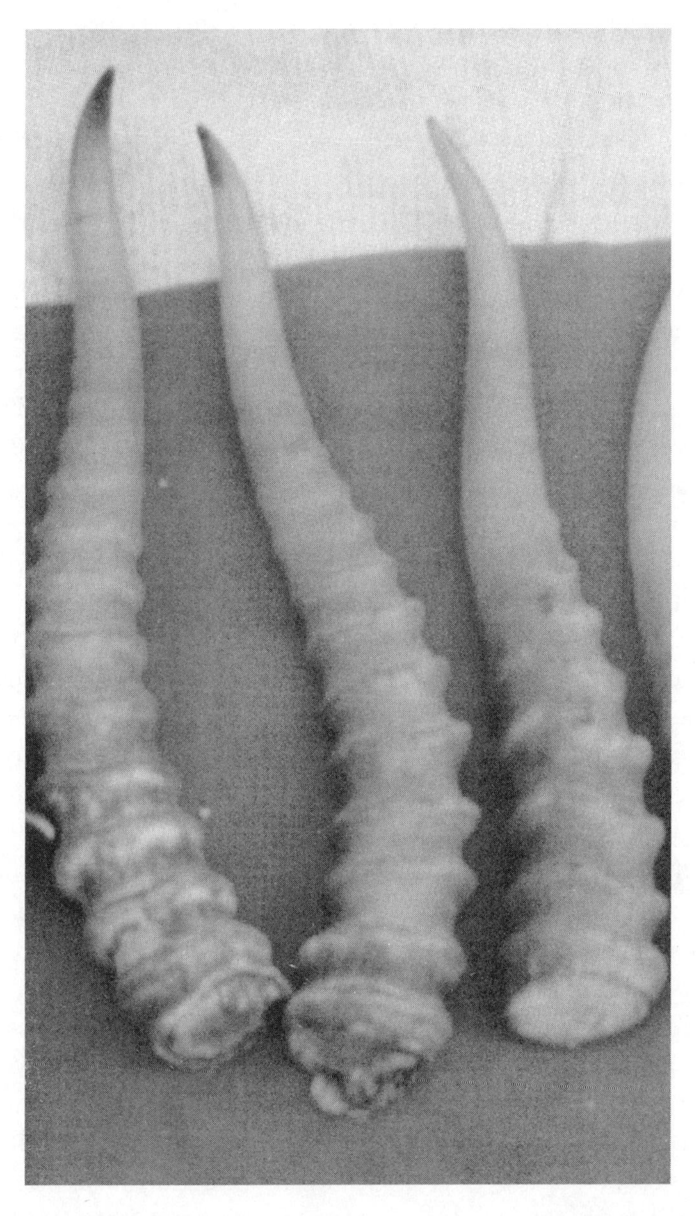

From the cover of the 1994 TRAFFIC report on the trade in saiga horns.
The report found that saiga antelope were being harvested unsustainably,
driven by the demand for traditional Chinese Medicine.

The saiga is not mentioned in the Chinese materia medica of 1597 (*Pen Ts'ao Kang Mu*), because at that time, according to a 1995 TRAFFIC report by Chan et al., "the real Saiga Antelope and its horn [were] largely unknown to the Chinese, let alone utilized in any way . . . there is no historical record of any Saiga horn trade between China and Central Asia along the Silk Road . . . [which] would have been the most likely route for trade in Saiga horn." By 1989, however, in Zhang Enquin's *Rare Chinese Materia Medica*, an entire section is devoted to *Lingyangjiao*, the horn of *Saiga tatarica*. We are warned to be on the lookout for counterfeits, such as the horn of the Mongolian gazelle (*Procapra subgutturosa*) or the Tibetan antelope (*Pantholops hodgsoni*), which might be processed to simulate the horn of the saiga. "After being soaked, dried, and ground to a powder, the horn can be used to check hyperactivity of the liver and relieve convulsion, treat the up-stirred liver wind, infantile convulsion and epilepsy; calm the liver and suppress hyperactivity of the liver-*yang*; it is efficacious in the treatment of dazzle and vertigo due to hyperactivity of the liver-*yang*; it improves acuity of vision, cures headache and conjunctival congestion; clears away heat and toxic material; and can be used to treat unconsciousness, delirium and mania in the course of epidemic febrile diseases." Like rhino horn, saiga horn is classified as a product "salty-cold in character and which can detoxify the body and reduce 'heat'" (Chan et al. 1995).

It did not require much effort to eliminate the unicorn from the earth—it never really existed in the first place—but some other horned animals, such as rhinos and saigas, have fared badly at the hands of hunters who would fill the demands of traditional Chinese pharmacological suppliers. The attempt to substitute saiga horn for rhino horn hasn't saved many rhinos and clearly has had tragic results for the saiga. Where Western medicine prescribes synthetic compounds such as aspirin and ibuprofen for reduction of swelling, pain relief, and headache—and has even developed a synthetic stimulant (Viagra) for erectile dysfunction in men—TCM cleaves to natural drugs, massage, and acupuncture. If and where the drugs are available, many people can get the same results with aspirin and ibuprofen, but believers in TCM prefer genuine rhino horn, and with the collaboration of the suppliers, species are being sacrificed in the process. But a certain hornless animal, celebrated for its majesty, vitality, and beauty, is scheduled to join the rhinos on the short list of animals threatened with extinction in the wild by the requirements of traditional Chinese medicine.

5

Where Have All the Tigers Gone?

Know also that the Great Khan has many leopards which are good for hunting and the taking of beasts. . . . He has several great lions, larger than those of Babylonia. They have very handsome, richly colored fur, with longitudinal stripes of black, orange, and white. They are trained to hunt wild boars and bulls, bears, wild asses, stags, roebuck, and other game.

Marco Polo, 1298

Around AD 70, in his *Historia Naturalis*, the Roman writer Gaius Plinius Secundus, known as Pliny the Elder, wrote this about the tiger (translation by Philemon Holland, 1601):

TYGERS are bred in Hircania [that part of Iran on the coast of the Caspian Sea] and India: this beast is most dreadfull for incomparable swiftnesse, and most of all seen it is in the taking of her young: for her litter (whereof there is a great number) by the hunters is stolen and carried away at once, upon a most swift horse for the purpose. . . . But when the Tigresse commeth and finds her nest and den emptie (for the male Tigre hath no care nor regard at all of the young) she runnes on end after her young ones, and followeth those that carried them away, by the scent of their horse footing. They perceiving the Tigresse to approach by the noise that shee maketh, let fall or cast from them one of her whelpes: up shee taketh it in her mouth, and away she runneth toward her den swifter, for the burden that shee carrieth: and presently she setteth out againe, followeth

> the quest after her fawnes; and overtaketh the hunter that had them away. Thus runneth she too and fro, untill she see that they be embarked and gone, and then for anger that she hath not sped of her purpose, she rageth upon the shore and the sands, for the losse of her fawnes.

Tigers probably became known to the ancient Greeks through Alexander the Great's campaigns in India and Persia, and they were occasionally used in Roman arenas. "By far the greatest number ever seen at one and the same time," wrote C. A. W. Guggisberg (1975), "were fifty-one, all of them massacred in the course of the games Heliogabalus arranged to celebrate his marriage. The same emperor is said to have harnessed tigers to a chariot on which he himself posed in the guise of Bacchus." After the heady days of the Romans, tigers seemed to have disappeared from European consciousness until the reappearance of Pliny's texts in the form of the *Physiologus*. "After the disintegration of the Roman empire, no tigers were seen in Europe for a very long time," Guggisberg tells us, "and the memory of the creature faded away so completely that Marco Polo was greatly puzzled by the 'lions' he saw at the court of Kublai Kahn. . . . The first tiger seen in Europe since Roman times seems to have been a specimen obtained by Yolanda, Duchess of Savoy, in 1748, and kept at her castle in Turin."

A half a world away, tigers had been an integral part of life (and death) for centuries. In India and Southeast Asia, they roamed the forests and woodlands, often preying on livestock and occasionally attacking and eating people. Sealstones from the five-thousand-year-old ancient civilizations at Mohenjo Daro and Harappa depict tigers, sometimes in captivity. In Hindu mythology, Durga is a fierce form of Devi, the Mother Goddess, depicted as extremely beautiful and full of rage, almost always shown riding a tiger. Shiva, the god of destruction, was once attacked by a tiger, but he caught the animal with one hand, peeled off its skin, and slung it over his shoulder like a shawl. At a more mundane level, Indian princes regarded tigers as fitting subjects for royal hunts, and often the great cats were herded by hundreds of beaters who crashed noisily through the brush until they drove the tigers into rifle range of maharajas, perched safely on the backs of elephants. Vast tracts of Mughal India were reserved as hunting preserves for royal tiger hunts. Of the hunts of Maharaja Ramsinghji of Kotah in 1771, Kailash Sankhala (1977) has written: "Every tiger hunt of that time was celebrated with feasting, dancing, and music; even the royal horses were included in the lavish feast given by the Maharajah celebrating the killing of a tiger." The record for tigers killed by one man in that era belongs to the Maharaja of Surguja, who claimed to have killed 1,150.

Tipu (also "Tippoo") Sultan, known as "The Tiger of India"—who was defeated at Seringapatam in 1792 by Lord Cornwallis (the general who had surrendered to George Washington in 1781) and then again in 1799 by Richard Wellesley, the brother of Arthur Wellesley, the future Duke of Wellington—was particularly obsessed with the tiger image. Stylized tiger stripes (*bubri*) are everywhere present—embroidered, quilted, painted, incised, and even forged into the watered steel of sword blades, while his banner was emblazoned with the words, "Tiger is God." His dreadful toy, the tiger organ, which showed a model tiger mauling a British soldier, is now in London's Victoria and Albert Museum. At the fall of Seringapatam, the British issued a commemorative medal showing the British lion overpowering the Indian tiger. (In later years, British sportsmen would further overpower Indian tigers by shooting them.)

Not as large as the rhino, but much more beautiful and also in a precipitous decline, the tiger represents the quintessence of raw energy and power. It was these qualities that made the tiger the object of centuries of "big game hunting," and now, not only can these qualities not protect it, but they are the very attributes that make parts of the tiger—almost any parts—highly desirable to human beings who believe they can acquire the tiger's wild energy by consuming the essence of the great cat.

Venerated for their strength, grace, and above all their beauty, tigers are celebrated in the art of every nation, especially those countries where tigers formed part of the landscape and history. Uncountable are the representations of the tiger in the arts of India, Pakistan, Nepal, Myanmar, Laos, Thailand, Cambodia, and Indonesia. There are few tigers left in China today, but they have appeared regularly throughout the three-thousand-year art history of that country, and every twelve years in the Chinese calendar is the "Year of the Tiger," a fact not left unnoticed by the forty-five countries and regions that issued "Year of the Tiger" stamps in 1998, including, of course, China itself. Beyond China, tiger images are everywhere, from the names of sports teams and nicknames of golfers to petroleum company slogans. The Tamil rebels in Sri Lanka call themselves the Tigers, and, at least in English, a person of high sexual energy is a tiger. Former United States Secretary of State George P. Schultz is said to have a tattoo of a tiger on his derriere in tribute to Princeton, his alma mater. The dragon and the tiger share equally important positions in the mysterious pseudoscience known as *feng shui*, and it was convenient for pharmacological purposes that at least one of these animals was available.

In late 2004, the satellite television channel Animal Planet conducted a two-month survey in seventy-three countries to identify the world's most popular animal. Viewers were offered a short list of ten animals. After more than fifty-two thousand votes were counted, the tiger

This life-sized hand organ, operated by turning a handle on the tiger's left side, was made for Tipu Sultan in France. Air is pumped into the bellows within the tiger's body and expelled as a shriek and a loud roar. As the tiger "kills" the British soldier, the victim's hand moves up and down, and tunes could be played on the button keys in the tiger's side.

was declared the world's favorite animal. (The dog was second, then the dolphin, horse, and lion, followed by the snake, elephant, chimpanzee, orangutan, and whale.)

In a 1999 essay entitled "The Tiger in Human Consciousness and Its Significance in Crafting Solutions for Tiger Conservation," Peter Jackson explains the importance of tiger symbolism in many Asian countries:

> At the advent of the Year of the Tiger these markings are painted on the foreheads of children to promote vigour and health. The children also wear tiger caps and tiger slippers for the occasion. In Korea, the tiger became the symbol of the Mountain Spirit, and the White Tiger the Guardian of the West. In modern times, South Korea chose the tiger as the symbol of the 1988 Olympic Games in Seoul. Malaysia has two tigers supporting its National Crest. . . . India, home of half the world's surviving tigers, replaced the lion with the tiger as its national animal when conservation started in the early 1970s. In Pakistan, where no tigers are now found, the political party which fought for independence from British rule, the Muslim League, still has the tiger as its election symbol and displayed it during its victorious campaign in 1997. Bangladesh has the image of the tiger on its banknotes. Because the tiger is the symbol of power, Hong Kong, Malaysia, Singapore, South Korea, and Thailand have been dubbed "Asian tigers" because of their rapid economic advance.

In the movies, the tiger is always portrayed as a powerful, dangerous creature, usually a man-eater. In the 1942 adaptation of Kipling's *Jungle Book*, Shere Khan the tiger is described as "villain, killer, man-eater, assassin," and establishes his bona fides early in the film when he attacks and kills a poor villager. Later, Mowgli is told (by Kaa, the python) to lure Shere Khan into the water so he can kill him "because tigers, like all cats, fear water," and he does just that. (Tigers do not fear water, but the producers' understanding of natural history was a little shaky, anyway, for among the monkeys shown in the Indian "jungle" are spider monkeys, found only in South America; and when Mowgli is shown in the den of the wolves that raised him, the "wolf pups" are little Labrador retrievers.) Tigers also appear as roaring, dangerous killers in the 2000 film *Gladiator*, where they are released into the Coliseum to attack the eponymous fighters, but a mere tiger—especially one on a chain—is no match for the ex-general Maximus, and he handily kills one with his sword. (How the heavy-coated Siberian tigers were supposed to have been brought to ancient Rome is not explained.) The *Last Samurai*, made in 2003, begins with a montage showing how the Japanese warriors trained: they practiced their swordsmanship on a snarling white tiger.

There are no tigers in Japan now, and there have not been since the late Pleistocene, according to Andrew Kitchener (1999), but their absence from the landscape has not precluded their appearance in Japanese art. There are many examples of tigers carved as netsukes, miniature sculptures that are now highly desirable collector's items. Carved of ivory, wood, or semiprecious stones, they represent a unique form of sculpture that can be held in the palm of the hand. Traditionally, women kept small objects in their sleeves, but because kimonos had no pockets, men carried *sagemono*, silk bags that were used to hold tobacco pouches, pipes, purses, or writing implements. The sagemono was hung from the kimono sash (obi) by a silken cord, and the netsuke served as the toggle that kept the cord from sliding through the sash.

Valmik Thapar, one of the world's leading tiger conservationists and author of many books and papers on tigers, presents in *The Tiger's Destiny* (1992) a fine discussion of the tiger's place in Taoism and Chinese culture:

> Taoism believes everything has a soul, be it animate or inanimate. When it is good, it is controlled by yang or the green dragon; when evil by yin or the white tiger. The breath of the tiger creates the wind, the breath of the dragon the clouds, and together they produce torrents of rain which regenerate the earth and provide vital food for man. In times of severe drought, real tiger bones would be dropped into a "dragon well," apparently causing such havoc for the reigning dragon that a vast storm would engulf the land, bringing endless rain and relieving the drought.

> The founder of Taoism, Chang Tao-Ling, was dedicated to the search for a dragon-tiger elixir which would grant eternal life. . . . His recipe for immortality is secret and exists only in a few mysterious writings. But it is significant that Chang is always depicted riding a tiger.

More than 2 million years ago, *Felis paleosinensis*—"ancient cat from China"—roamed the forests of central Asia, spreading westward to the shores of Caspian Sea and the Caucasus and into Persia and Afghanistan; early tigers also spread eastward into Manchuria and Korea. For the most part, tigers from northern climes grew larger than those from farther south. (In modern times, the largest tigers are the "Siberian" forms; the smallest were the extinct Balinese tigers.) From Southeast Asia and India they reached the islands of Sumatra, Java, and Bali. Curiously, although tigers are excellent swimmers and must have colonized the Indonesian islands by swimming there, they never managed to cross the Palk Strait from southern India, and there are not now, nor have there ever been, tigers in Sri Lanka.

Saber-toothed "tigers," the large cats that lived in Europe and North America during the Pleistocene, some one hundred thousand years ago, are not included in the ancestry of modern tigers. In *The Big Cats and Their Fossil Relatives* (1997), Alan Turner explains:

> While the term "saber-tooth" (from *saber*, a curved sword often used by mounted cavalry) describes some features of the dentition of these cats well enough, there is simply no basis for calling them tigers; they are not closely related to the true tiger and are certainly not an ancestor of that species, and there is no evident reason to assume that they had striped coats.

There were many species of early cats, some of which were as large or larger than modern tigers, and many of which had characteristically enlarged canine teeth. Because of small but significant differences between many of the earlier species and modern cats (Felidae), they have been classified separately as Nimravidae. Some nimravids, such as the leopard-sized *Hoplophoneus occidentalis*, had canines that were only moderately enlarged, but *Barbourofelis fricki*, which was larger than a modern lion, had huge stab-

The saber-toothed "tiger", unrelated to today's tigers, became extinct about fifteen thousand years ago.

bing teeth that fit into a flange on the lower mandible. Of the many bones of large cats that have been recovered from the La Brea Tar Pits of Southern California—according to Christopher Shaw (1992), the number exceeds 160,000—probably the best known are those of the saber-toothed cat called *Smilodon fatalis*, described by Shaw as

> about the same size as an African lion; however the body proportions were different from those of any large living cat. Its robust front limbs indicate that it sought prey that greatly exceeded its own weight, but its skeletal structure suggests that it did not chase its prey extensively. This animal was an active predator that relied on stealth and ambush rather than speed, using surprise and a short, rapid pursuit, followed by a violent impact and a lethal bite.

This is essentially a description of the modern tiger's hunting technique, but while its forelegs are powerful enough to break the neck of a deer with a single blow, they are not proportionally as long as those of

Smilodon. Tigers are unmitigated carnivores; they prefer large prey to smaller but will hunt the smaller if necessary. According to John Seidensticker, in his 1996 book *Tiger*, "Tigers specialize in killing large deer, wild cattle, and wild pigs. . . . Tigers can kill prey ranging in size from the large males of the wild cattle species, which top the scales at 1,000 kg (about 2,200 lb), to the diminutive muntjacs weighing just 15 kg (33 lb)." They stalk their prey, usually in thick cover, keeping low to the ground and inching forward until they are in range, then charging with astonishing speed. The tiger's preferred method of killing is a powerful bite to the throat, but struggling deer or buffalo do not always cooperate, and often the tiger has to wrestle the animal to the ground before administering the windpipe-crushing bite. Smaller deer, such as the chital, may be killed with a single forearm blow to the head, but tigers have to puncture the throat of the larger, heavier sambars. Many times the tiger is spotted before its charge, or the charge spooks the animal, or even when contact is made, the prey animal gets away. The general wariness of tigers and the deep cover in which they live has made it almost impossible to observe the tiger's hunt, but it has been estimated that only 10 percent of the charges are successful.

With the exception of lions, which live in communal groups known as prides, and cheetahs, where adult females stick with their dependent cubs, all other wild cats are loners, none more so than the tiger. Male and female must come together for mating, but when that has been accomplished, the male moves off and may never see his offspring again. The female remains a solo act until her cubs are born, and after nursing them for six to eight weeks, she hunts for solid food for herself and for them. By the age of six months, the cubs are completely weaned, and males—which grow larger and faster than females—will weigh about 100 pounds, while the females will be about 30 pounds lighter. At the age of about eighteen months the cubs get their permanent canine teeth and are ready to make their own kills. At this age, males begin to wander off in preparation for establishing their own territory. Males and females stake out their own territories, but that of the male covers much more acreage. Tigers are so dedicated to the concept of territoriality that even when baits are regularly put out at particular locations, the tigers continue to make their regular rounds, as Charles McDougal, the chief naturalist at Tiger Tops in Chitwan, points out in *The Face of the Tiger*.

Because tigers rarely see others of their species, they rely on scent marking to keep aware of other tigers. With skin glands around their cheeks, toes, tail, and anogenital areas, they leave scent marks by rubbing these parts on trees and bushes, and also leave "signatures" in their urine and feces. As tiger researcher K. Ullas Karanth (2001) has written, "When we see a tiger, we may think it is walking alone, although it is leaving scent messages for other tigers almost continuously. The tiger may be solitary, but it is not alone." In *The Secret Life of the Tigers*, Valmik Thapar, writing of the tigers of Ranthambhore, also suggests that tigers are not always as solitary as sometimes portrayed. He tells of a male tiger who remained and provided food for his cubs and their mother. And he "observed instances of communal feeding, where nine tigers shared a carcass, controlled by the tigress Padmini. In fact, six of these tigers were cubs of Padmini's litters . . . the first sign that tigers might sustain kin links throughout their lives." He even found a situation where a male tiger—probably the father—remains with the tigress and the cubs: "There was no question then, of the male tiger practicing infanticide . . . we now have evidence that in the course of the next few months the male took an active role in providing food for the cubs and their mother, and therefore had a vital role to play in raising the family."

Tigers are generally nocturnal but have been observed hunting in the cool of the morning and evening as well. Like all cats, tigers see well in limited light, and, like lions, leopards, and jaguars, in bright sunlight their pupils contract into small round shapes, not the vertical slits of the smaller cats. Roaring is most often associated with lions, but tigers, jaguars, and leopards (but not snow leopards) also produce these hollow, guttural sounds. It is believed that roaring functions as some sort of long-range communication for tigers, but its actual purpose—unless it is merely to advertise the cat's presence—is not fully understood. It is a sound to be reckoned with, as Richard Perry noted: "The full-throated roaring of a tigress, or tiger, rolling along the earth in deep and thrilling gusts of sound to a distance of two miles or more on still nights, upsets all the jungle . . . as that solemn and vibrating *aa-oo-ungh!*, as if from the depths of a great organ, reverberates down the aisles of the valleys and over the jungle; terminating after a brief pause in a series of explosive booming *oo-oo-oonghs!*"

A recent research project by acoustician Edward Walsh and his colleagues at the Developmental Auditory Physiology Laboratory in Omaha, Nebraska, suggests that tigers use low-frequency sounds—some so low as to be inaudible to humans—to communicate with one another. Working with tigers at the city's zoo, the researchers found that some of the tigers' sounds "stretched down into the infrasound range, below 20 hertz" (Anon. 2003a).

Although their eyesight and hearing are incredibly keen, tigers are said to have a very poor sense of smell. In his 1967 landmark study, *The Deer and the Tiger*, George Schaller wrote, "The role of olfaction in the communication between tigers provides an interesting sidelight on a hotly debated issue among hunters. The tiger's sense of smell is usually said to be either poor or practically absent. . . . [But] the fact that the tiger uses scent as a means of signaling certainly indicates that its powers of olfaction are good." Given the acuity of its other senses, it would give the tiger an unfair advantage if it had a sense of smell that was equal to its sight or hearing. If the tiger's olfactory ability is controversial, there is no question about the sense of the prey animals. To detect and avoid predators such as the tiger, hoofed animals usually have well-developed eyesight, hearing, and a sense of smell. As Peter Matthiessen wrote, tigers "have never adapted to their own strong smell by learning to hunt upwind when stalking—one reason they miss at least nine kills out of ten."

Although tigers are hard to find and therefore notoriously difficult to count, the worldwide total of all species and subspecies in the wild is generally thought to range between 2,500 and 4,000. There is no way of telling how accurate these numbers are, and it is possible that the number is even lower than the low estimate. In his 1996 book *Of Tigers and Men*, Richard Ives claimed that only 700 tigers of any species were left in the world, but hardly anyone—no matter where they stand on the issue of tiger conservation—agrees with him.

Although they are all classified as *Panthera tigris*, two forms are usually recognized: the larger, lighter-colored, heavier-coated "Amur" tiger (known by the subspecies name of *Panthera tigris altaica*) and the somewhat smaller, darker, and slenderer "Bengal" tiger. All wild tigers are threatened by traditional Chinese medicine, and of the eight recognized tiger subspecies, three have already been exterminated. The living five are the Amur (Siberian) tiger (*Panthera tigris altaica*), the Bengal tiger (*P. t. tigris*), the Indochinese tiger (*P. t. corbetti*), the South China tiger (*P. t. amoyensis*), and the Sumatran tiger (*P. t. sumatrae*). The Bali tiger (*P. t. balica*) was the first to go—the last confirmed sighting was in 1939—followed by the Caspian tiger (*P. t. virgata*), last seen alive in 1968, and then the Javan tiger (*P. t. sondaica*), which has not been spotted since 1976.

For convenience, *Panthera tigris* has been divided into eight subspecies, but the distinctions are often variable and there are many biologists who regard them as being arbitrary and meaningless. Some distinctions are based on size, some on the ground color of the coat, and some on stripe color and striping patterns, but these may vary widely within a given population and within a given geographical area. In his 1999 chapter, "Tiger Distribution, Phenotypic Variation and Conservation Issues," Andrew Kitchener pointed out that "striping patterns and coloration attributed typically to Javan and Bali tigers were also found in animals from Sumatra, Burma, and South India." And in coloration, "variation within the putative subspecies may be greater than variation between them." After careful consideration of the variables, Kitchener concluded that there are only two subspecies: *P. tigris tigris*, of mainland Asia, and *P. t. sondaica* of the Sunda Islands and possibly peninsular Malaysia. But even if these two are the only valid subspecies, so much has been written as if there were eight that in this discussion I will recognize the eight nominal subspecies especially when quoting authors who believed they were writing about a subspecific population or individual.

Neither a species nor a subspecies, white tigers have always been considered very special. Probably based on an occasional sighting in the wild, they were regarded as particularly sacred and therefore formed an integral part of Chinese ritual. The white tiger (*Pai Hu*) was believed to control wind and the water, and its image was often placed on tombs and temples. White tigers may make for spectacular zoo or circus exhibits—for their Las Vegas act, Siegfried and Roy have fifty-eight—but because they are both crossbred and inbred, they have no conservation significance and some regard them as an aberration of nature. At a

1986 tiger symposium held at the Minneapolis Zoo, an entire session was devoted to white tigers, and Edward Maruska of the Cincinnati Zoo presented a paper he called "White Tiger: Phantom or Freak?" Maruska identified our fascination with "strikingly colored black and white animals," such as the giant panda, killer whale, zebra, colobus monkey, and ruffed lemur, but then quotes William Conway of what was then called the New York Zoological Society (now the Wildlife Conservation Society) as saying, "White tigers are freaks. It's not the job of a zoo to show two-headed calves or white tigers." Lee Simmons of the Henry Doorly Zoo in Omaha, Nebraska, wrote that "you justify white tigers in the same way as you justify traveling giant pandas or koalas or any other high-visibility animals, which, through the ability to catch the public fancy, significantly enhances public support and therefore the financial well-being of your institution." In other words, your zoo needs money to carry on its good works, so you do whatever is necessary to keep the money coming in—even exhibiting two-headed calves.

We don't know how frequently white tigers appear in nature, but we do know that in the last one hundred years, only about a dozen have been seen in India. (White forms have never been reported for any of the other subspecies.) The precursor to all captive white tigers is believed to be a nine-month old cub that was trapped by hunters in the forests of Madhya Pradesh in 1951 and taken to the palace zoo of the Maharaja of Rewa at Govindgarh. Named Mohan, the cub was later mated to a normal-colored captive tigress who produced three litters with normal coloring. A few years later, Mohan mated with one of the offspring, producing the first litter of white cubs in captivity; they are the ancestors of others now in zoos the world over. White tigers in zoos today are inbred and crossbred mixtures of Bengal and Siberian tigers. They are neither albinos (in which case they would have pink eyes) nor a separate species; they have chocolate stripes and blue eyes, although several variations in eye and stripe color have been seen. White tigers are only born to parents that both carry the recessive gene for white coloring.

With the obvious exception of the white ones, all tigers are orange with black stripes and white undersides. This coloring makes the tiger stand out dramatically in a zoo environment, but in the wild, the vertical striping provides surprisingly good camouflage. (An orange and black animal in the snow, however, is hard to miss.) While the various subspecies differ slightly from one another, they all conform to the same general color scheme. Whether found in snowy forests or in steaming jungles, every species of tiger is being hunted down and killed for Chinese medicine.

The Siberian (Amur) Tiger

As David Prynn points out in the introduction to his 2004 book, *Amur Tiger*, "This book is about the Amur, Ussuri, or so-called Siberian tiger. The Amur and Ussuri rivers are within the tiger's historical range, [and] so are acceptable names; 'Siberian' is not recommended, because the tiger has probably never occurred in Siberia as normally understood in Russia."

The largest of the tigers—and the largest of all living cats—the Siberian (Amur) tiger, *Panthera tigris altaica*, is found primarily in the coniferous, scrub oak, and birch woodlands, not of Siberia, but of far-eastern Russia (Ussuriland and Khabarovsk Province in particular), with a few remaining in northeastern China. Because of its heavier coat, the Amur tiger appears much larger than the Indian and Southeast Asian races, but as Peter Matthiessen noted in *Tigers in the Snow*, "it is only two to four inches taller at the shoulder than those mighty Bengal tigers."

The Amur tiger needs large prey, such as boar and red deer, to survive and there are known cases of tigers killing and eating adult brown bears. The largest Amur tiger reliably recorded was a male shot in the basin of the upper course of the Sungari River in Manchuria in 1943. He was 3,507 millimeters (11 feet, 6 inches) in length, including tail. Males can weigh up to 660 pounds, though David Macdonald's *Encyclopedia of Mammals* suggests that one was found that weighed 384 kilograms (845 pounds). Females, in any case, are smaller, measuring about 8½ feet from head to tail and weighing 200 to 370 pounds. The individual territory of Amur tigers is also quite large, ranging from 39 to 154 square miles for females to an astounding 309 to 390 square miles for males.

In his 1994 history of the Russian Far East, John Stephan gives us a basic overview of the Amur tiger situation at that time:

Amur tiger at rest in Primorski Krai, in the Russian Far East.

Ginseng and tigers thrive in the Sikhote Alin Range. The former attracted people; the latter ate them, if one is to believe lore about peasants being pulled through izba windows and soldiers being snatched from marching columns. In point of fact, tigers disconcerted rather than devoured nineteenth-century settlers. Unaccustomed to sharing space with humans, they prowled around farms and occasionally made themselves at home in outdoor privies and bathhouses. Although buttons and a butterfly net were all that was left of an unlucky German lepidopterist in 1914, the fang-toothed felines were more partial to puppies. Dogs vanished from the streets of Vladivostok in the 1860s, guaranteeing a niche for their predators in municipal nomenclature (Tiger Street, Tiger Hill, etc.). Hunted to the verge of extinction by the 1950s, Ussuri tigers rebounded as a protected species, confirmed by paw prints and telltale canine remains in Vladivostok as late as 1986. In the early 1990s, however, their numbers again fell as poachers catered to the Chinese market for tiger skins.

When the Soviet Union collapsed in 1991, law and order collapsed with it. In a land where hunting was, and still is, a way of life, the value of tiger skin and bones for sale to China, the Koreas, Japan, and Taiwan had become an incentive to kill tigers. It was feared that that the tiger would be eliminated in a few years. An extensive survey of Amur tigers conducted in 1995–96 (Matyushkin et al. 1997) found that, as expected, the great majority lived in the Russian Far East, that perhaps 20 lived in adjacent regions of northwest China, and that there were only unconfirmed reports that Amur tigers remain on the Korean peninsula. More than 650 field workers conducted the survey over approximately 120,000 square miles, an area the size of Norway, mostly encompassing the Sikhote-Alin mountain range. In this vast area (and in winter weather) they found approximately 450 adults, subadults, and cubs. Expanding ties within China provided an easy avenue for moving tiger products through Vladivostok and Khabarovsk. Peter Matthiessen, in *Tigers in the Snow*, provides a complementary perspective:

> The depletion of China's hoard of tiger bones explains the upsurge of the trade in the late 1980s, when poaching intensified in India, Bhutan, Indochina, Indonesia and the Russian Far East. Many of the golden hides eventually made their way to the Arab states, while the bones went to dealers in China and Hong Kong, Taiwan and Korea, Singapore, Japan, and the large Asian communities abroad . . . . Tonics and potions brewed from genitalia were much in favor among rich, flagging Asians, and a dried penis (which resembles a coiled eel) could command up to $2,500 in Singapore or Taiwan, where bowls of nice hot penis soup, $300 each, were consumed for their restorative properties in man's struggle against such dire afflictions as impotence and death.*

In an effort to develop stronger conservation measures for Siberian tigers, Wildlife Conservation Society tiger specialist John Goodrich, in

* In his 1987 article "Deadly Love Potions" (deadly to the animals that provide the material for the potions), Esmond Bradley Martin tells us: "Elderly Chinese men regard the male sex organ as the best virility prescription of all. They place a dried tiger penis, with testicles still attached, into a bottle of French cognac or Chinese wine and let it soak for weeks, or even months. Then they take a few sips of the liquor every night."

a joint program with the Society's Hornocker Institute and the Sikhote-Alin State Biosphere Zapovednik, radio-collared more than thirty tigers and followed their movements year after year. Ten years on, Goodrich and his colleagues reported in the February 2002 issue of *Wildlife Conservation* that four of the five tigresses collared in the region of Koonalayka Creek had been killed by poachers and ten of twelve cubs died. Where there were no logging roads that allowed poachers easy access to the tigers' remote habitats, the tigers fared much better. Therefore, Goodrich and his colleagues suggested that roads no longer in use be destroyed or made impassable, which would restrict access for poachers and provide no easy way for them to get the carcasses out, especially if the region was then guarded.

In preparing *Monster of God*, his 2003 study of man-eating predators, David Quammen also traveled to the Sikhote-Alin mountains to talk to the people whose daily lives include occasional contact with the animal they call *amba*, the Amur tiger. Quammen hooked up with biologist Dimitri Pikunov ("a burly, impetuous man of vehement moods and quiet charms") and learned the complex history of the snow tigers in Russia. Until the end of the nineteenth century, some 150 were killed annually, mostly for the pelts, but also because they posed a threat to hunters and trappers. When the Politburo decided that the Russian Far East should be colonized around 1930, tigers presented a risk to the "colonists" (often forced laborers) and were eradicated. Tiger hunting was prohibited in 1947, but the live capture of cubs continued. When a census was conducted in 1969–70, some 250 tigers remained in the Russian Far East. Then the Chinese market for tiger parts opened up. Quammen estimated that at a rate of about sixty tiger deaths per year (the rate in 1993), the Amur tiger would last only a decade. For the Far East as a whole, timber sales could be worth far more economically, but for local villages in the short term, the difference that the money a tiger sale could make was huge.

In the *New York Times* on January 16, 2005, James Brooke described his visit with Dimitri Pikunov in Nezhino, near Vladivostok, and wrote about a new survey of Russian tigers to be conducted in February, 2005. As in the 1995–96 attempt, volunteers will fan out along 600 known tiger routes, checking for the distinctive paw prints in the snow. The

1995–96 count estimated 450 tigers, down from the 600 that had been estimated in a 1990 survey. "Since then," wrote Brooke:

> "there have been fears that the population has been diminished further. With the fall of Communism, tigers, along with other wildlife here, have been hurt by the collapse of Soviet controls on hunting and trade and an explosion in Chinese demand for wildlife delicacies and traditional medicines. Chinese buy frogs, bear paws, wild ginseng, deer antlers, and deer penises. . . . As evidenced by the dead tigress found in a Chinese snare near here [Nezhino] in November, the top cat regularly loses out to the top primate."

Pulling triggers, setting snares, rigging electrical wires, or poisoning water holes, poachers are also laying waste to the diminishing populations of tigers in India and Southeast Asia. In countries where tigers have been symbolically dominant for ages, the ultimate source of the symbol is fading.

The Tigers of Indonesia

> There are always a few tigers roaming about Singapore, and they kill on an average of a Chinaman every day. . . . We heard a tiger roar once or twice in the evening and it was rather nervous work hunting for insects among the fallen trunks and old sawpits, when one of the savage animals might be lurking close by, waiting an opportunity to spring upon us.
>
> *Alfred Russel Wallace*, *The Malay Archipelago*, *1869*

South of Borneo, Java is only the fifth largest of the Indonesian islands but it contains two-thirds of the country's population. Indeed, it is one of the most densely populated areas in the world, with more than 100 million people living in an area the size of Alabama. Jakarta (formerly known as Batavia), the capital of Indonesia, is located there, with a population of 8.25 million. After 1619, the Dutch East India Company's settlement at Batavia became the trading and shipping center for the "East Indies." Because of the importance of the center, the colonial government was particularly interested in ridding the region of dangerous animals.

Tigers troubled the people of Batavia from the very beginning of the Dutch presence, wrote Peter Boomgaard (in *Frontiers of Fear: Tigers and People in the Malay World, 1600–1950*); and although the governors themselves were rarely active in tiger hunting, fairly soon their role seems to have evolved to giving money to people who presented them with captured or killed tigers. Sometimes the rewards were equal to a man's rice ration for an entire year, so it is understandable that people would seek out tigers to kill, even if they weren't threatening anyone. Hunters either surrounded the animals and dispatched them with spears or dug pitfalls lined with sharpened bamboo stakes. They also employed "tiger-traps," which consisted of a wooden cage that contained a live goat, and a trapdoor that could be triggered when the tiger entered. "According to the Hungarian hunter Count Andrasy, who traveled over Java in 1849–50," wrote Boomgaard, "some 400 tigers annually were captured in these traps." Meanwhile, the sultans and rulers of various kingdoms on Java were hunting tigers—somewhat ineffectually, as elephants were not available to offer them protection—both to rid the countryside of these dangerous animals and to enable the aristocracy to demonstrate its prowess. And the Tuwa Buru people of Java paid taxes to their rulers in live tigers and deer. Thus the Javan royalty, Javan native peoples, and Dutch colonials in effect conspired to rid the island of tigers, and their efforts, not altogether successful in their lifetimes, certainly contributed to the downfall of *Panthera tigris sondaica*, the Javan tiger.

Java was also the scene of rituals, recorded as early as 1605, that were not specifically designed to reduce tiger populations but because the tigers died in the process, the end results were the same. One such ritual, reminiscent of the Roman stadium, involved placing a tiger and a "buffalo" (usually the wild ox known as the banteng) in an arena for a fight to the death. Contrary to popular opinion, tigers do not attack anything within range, and in most instances, the tiger had to be prodded to leave its cage, and the panicked "buffalo," which might have outweighed the tiger by a thousand pounds, often managed to kill the tiger. In the ritual known as *rampogan sima* ("tiger sticking"), caged tigers were released into an arena where they were surrounded by ranks of spearmen and desperately trying to find an escape route, were stabbed to death. According to Boomgaard, "such traditional rituals in which tigers were killed were strongly supported by the colonial state and even 'invented' in areas where they had not been carried out before. Together, the colonial state and the European hunter were making the Orient safe for the Empire." The ritual practices faded from view around 1900, probably because tigers were becoming too scarce.

As forested areas, alluvial plains, and river basins were converted to agriculture, the Javan tiger's habitat was concurrently constricted, and because the agriculturalists regarded tigers as a dangerous nuisance, many were poisoned. By 1970, Java's last tigers were confined to the wild southeast coast known as Meru-Betiri, but because their prey—the small sambar deer known as the rusa (*Cervus timorensis*)—had been decimated by disease, there was nothing for the remaining tigers to eat. In a paper poignantly titled "Bearing Witness: Observations on the Extinction of *Panthera tigris balica* and *Panthera tigris sondaica*," John Seidensticker reported that in 1976 he and his colleagues had found tracks of three animals in the Meru-Betiri National Park, but after 1979 they found no evidence to indicate that any tigers were still alive in the region.

Throughout history, Javan tigers have been killed by royalty, native peoples, and then Dutch colonials, but during the nineteenth century, big game hunters took up the slaughter. According to Andries Hoogerwerf, author of *Udjung Kulon: The Land of the Last Javan Rhinoceros*, hunters competed to see who could kill the most: "The fanatical tiger hunter, A. J. M. Ledeboer, was out to beat the record of the Sumatran tiger hunter Hofman, who is said to have killed over 100 of these animals in Sumatra. Ledeboer succeeded in this sinister purpose as far back as 1933." Javan tigers were more often killed by eating poisoned wild boar, though. Hoogerwerf suggests, "The poisoning en masse of the wild boar, whose flesh may not be eaten by the mainly Moslem population of Java on the strength of religious considerations, has been going on for almost half a century and to an ever increasing extent, even being encouraged by the Government. It was perhaps the most important direct cause of the rapidly shrinking stock of the Javan tiger."

The hunt for tigers in Bali reached its height later, but was no less conclusive. Bali, Indonesia's prime tourist destination, is a much smaller island than Java, covering only 2,200 square miles. It is largely mountainous and now mostly under cultivation for rice, vegetables, fruits, and coconuts. Before rice and tourists, however, Bali was heavily forested and was home to a variety of wildlife, including deer, leopards, and tigers—the smallest and darkest tigers known. Bali is only a mile across the Bali Strait from Java, and because tigers are such good swimmers, it is likely that tigers occasionally crossed from one island to the other. Calculating one adult per 40 square kilometers, Seidensticker estimated that at most there were 125 tigers living on the island. Overwhelmed by human activities, the Bali tiger never had a chance and had disappeared by 1960 (Seidensticker, Christie, and Jackson 1999b).

After Borneo, which in square miles is a little smaller than Spain, Sumatra is the second-largest of the Indonesian islands. Most of the eastern half of the island is swampland, and much of the interior is impenetrable jungle. Like other islands nearby, Sumatra was long a part of the Dutch East Indies and was ruled by colonial governors from 1602 until 1798. Tigers represented a nuisance to the Dutch spice planters there, and beginning as early as 1838, bounties were offered for each tiger captured or killed.

From 1972 to 1975, Markus Borner traveled throughout the island of Sumatra, spending roughly two months in each of the eight provinces, trying to assess the size of the surviving tiger population. On behalf of the World Wildlife Fund's "Operation Tiger," Borner conducted numerous interviews, and based on the information he received, he selected promising survey areas and made expeditions to search for evidence—"scratching and pug marks, feces, remains of kills, and so on" (Borner 1978). He managed to find tigers in all eight provinces, but was the first to admit that his figures could only be an approximation, given the difficulties of surveying such an elusive animal in such rough terrain. Nevertheless, he published this summary:

Province	*Approximate Number of Tigers*
Aceh	100
North Sumatra	50
Riau	300
West Sumatra	50
Jambi	200
South Sumatra	100

Begkulu	100
Province	*Approximate Number of Tigers*
Lampung	50
Total	1,000*

* Borner's numbers actually add up to 950, but in his paper the total is shown as 1,000. Others who refer to Borner's table also use his total of 1,000, so I have used Borner's original total, even though it is obviously incorrect.

Borner recognized habitat destruction and hunting as the major threats to Sumatran tigers. A law protecting tigers seemed to have little effect, perhaps in part because there was continued killing of Sumatran people by Sumatran tigers. Borner wrote:

> Many horror stories circulate among the people of Sumatra describing tigers who have killed dozens of people. Villagers claim that in 1972 a tiger killed 30 people in a rubber plantation near Batu Radja (South Sumatra) . . . but I myself could confirm only three cases of human beings being killed by tigers.

Sumatran tigers are predominantly hunted today using wire snares, which are designed either to catch the tiger's leg or encircle the animal's neck. Poison, the Dutch colonial government's method of choice in the early 1900s, is still being used in limited quantities, and occasionally armed policemen or soldiers will shoot at a tiger when one appears. Even where they are supposedly protected, the remaining Sumatran tigers are poached for body parts, poisoned by villagers who consider the big cats a menace, and forced onto ever-smaller tracts of habitat due to logging in the forests. On May 19, 2002, the tables were turned, but not in a way that was likely to ensure the tigers' protection. A tiger (or tigers) killed two loggers in the forest of Jambi province.

Sumatran tigers now live in scattered populations in the island's deep forests, and the tiger's secretive nature means that few are ever seen. The forests that provide sanctuary for the tigers may eventually prove their undoing, for it is the clearing of these very forests at a very high rate that is the greatest threat. Since Borner's survey, the population of Sumatran tigers has continued to fall, standing at five hundred

Tiger shot in 1935 by hunters in Alahan Panjang, Sumatra. The population of Sumatran tigers was in decline even before the escalation of killing for the parts to be used in traditional Chinese medicine.

in 2003, according to Matthew Linkie. From 1996 to 2001, Linkie and his colleagues conducted field surveys in the 824-square-mile Kerinci Seblat National Park in Sumatra, recording tiger signs (pugmarks, scats, sightings) in 126 locations. For 5,500 photo-trapping events (in which an automatic camera had been set along known tiger paths), they recorded only thirteen adult tigers—seven males and six females. The tigers of Sumatra, whether you call them *Panthera tigris sumatrae* or *Panthera tigris tigris*, are critically endangered.

In the comprehensive 2004 TRAFFIC East Asia report *Nowhere to Hide: The Trade in Sumatran Tiger*, authors Chris R. Shepherd and Nolan Magnus estimated the number of wild tigers on Sumatra as probably fewer than 400, living in six protected areas. Another 100 or fewer tigers outside the protected areas probably will not last long. The six protected areas are the national parks Gunung Leuser (with an estimated 110–180 tigers), Kerinci Seblat (76–170), Bukit Tigapuluh (36), Berbak (50), Bukit Barisan Selatan (40–43), and Way Kambas (36). Factoring in the high estimates, that is a total of only 479 Sumatran tigers. In other words, more than half the number of Sumatran tigers that Borner estimated in 1978 were gone by 2004.

Indonesia passed a Conservation Act in 1990 designed to protect the tigers of Sumatra, but still the killing continues, for trophy hunting, protection of farmers and herdsmen, and, of course, for the needs of TCM. Campbell Plowden and David Bowles (1997) investigated the sale of tiger products in northern Sumatra, particularly around the city of Medan, because a large Chinese population resides there and also because of the proximity of Gunung Leuser National Park, with its relatively large population of tigers. (Gunung Leuser also has—or had—a large population of orangutans and is one of the last strongholds of the Sumatran rhino.) The investigators visited eighty-eight shops in and around Medan and found ten that offered verified tiger products for sale, but they also saw many bones, teeth, pelts, and skeletons. They did not find "evidence showing that there is organized poaching for tigers for trophy mounts or for the international trade in bones used in oriental medicine." But, they wrote, "it is apparent that many tigers that are killed opportunistically or deliberately by farmers are being fed into a commercial domestic market for tiger bones, teeth, claws and skins." To complete the circle, in her 2000 TRAFFIC report, Kristin Nowell wrote, "Investigation of Customs records in South Korea revealed that hundreds of kilograms of Tiger bones had been illegally imported in the years leading up to 1993. Countries of origin for the bones were listed, indicating that Sumatra was the major source."

Until around 1990, the most valued part of the Sumatran tiger was the skin. In 1985, according to Charles Santiapillai and Widodo Ramono, the retail price of a Sumatran tiger skin was US$3,000. But when the medicinal demand for tiger bone began to rise, the skins, while still valuable, were no longer the primary reason for tiger poaching. Indonesia became a primary exporter of tiger bone and tiger bone products, and while the primary recipient of these exports was South Korea, sizable exports were also recorded to Singapore and Taiwan. In Sumatra, tiger teeth, claws, whiskers, tail, pieces of skin, fat, and dung are also believed to have medicinal properties, while the penis and bone are believed to be aphrodisiacs. The TRAFFIC report concludes:

> Data collected by this survey indicates that Sumatran tigers are being killed and removed at an average rate of at least 51 tigers per year over the past five years. With a total population estimated at 400–500 tigers, this implies that at least 10% are being lost every year . . . for many reasons, it is likely to be an undercount, and annual losses are likely to be greater. . . . The survey indicates that poaching for trade is responsible for the vast majority (over 78%) of estimated tiger deaths.

The Tigers of China

In the past, wild tigers ranged throughout most areas of China and Manchuria that were not desert or high mountains, but for the most part, they are now gone. Tigers are an important element in Chinese iconography and, as we shall see, an even more important element in Chinese medicine. It was convenient to have tigers available in China, but this very convenience was responsible for the Chinese tiger's downfall and the subsequent need to seek tiger parts elsewhere.

The South China tiger (*P. t. amoyensis*), found in central and eastern China, is one of the smallest tiger subspecies. Males measure only about 8 feet from head to tail and weigh approximately 330 pounds. Females are even smaller, measuring about 7½ feet long, and when full grown weigh approximately 240 pounds, about the weight of a large jaguar. The short, broad stripes of the Chinese tiger are spaced far apart compared with those of Bengal and Amur tigers. During the 1950s and 1960s, the government of the People's Republic of China declared tigers a nuisance and sponsored a program designed to eradicate them in the country, rather like the campaign to eradicate flies in Beijing. Under this program, thousands of tigers were shot, bringing the South China tiger close to the brink of extinction.

In 1977 the Chinese government belatedly awoke to the fact that the tiger population was decreasing alarmingly, and in an effort to stop the decline, a law was passed forbidding the hunting of all tigers. Unfortunately, this law could not be strictly enforced and hunting continued. Now poaching is the principal cause of the tigers' decline. Poachers have stalked the animal so mercilessly that it has retreated deep into its natural habitat, and the black market continues to thrive.

At the 1986 Minneapolis Zoo tiger conference, Chinese zoo directors Xiang Peilon (Chongqing Zoo), Tan Bangjie (Beijing Zoo), and Jia Xianggang of the Division of Nature Conservations summarized the gravity of the situation at the time:

> Before the 1950s, the range of the South China tiger had been widespread throughout the vast territory south of the Changjiang (Yangtze) Valley; historically the provinces of Hunan and Jiangxi had been the centers of its distribution. . . . Later on, owing to the higher density of human population, more extensive and economic and agricultural exploitation, continued reduction of the forested area, and belated natural conservation, there was a sharp decline of population and a great reduction in range since the end of the 1950s. The present population of South China tigers, according to our preliminary estimate, is approximately 40.

The directors identified three ways of rescuing the South China tiger: (1) introduce new conservation laws, (2) protect the known wild tigers and their habitat, and (3) create new reserves. But, they said, "the most urgent, as well as effective, measure at the present time is to set up as quickly as possible a world central captive breeding group for the South China tiger." This was presented in 1986 and published the following year, just about the time the hunting of tigers for traditional Chinese medicine was beginning. The authors may not have been aware of the increase in poaching in China, India, and wherever tigers could be found, but as Chinese zoologists, they could not avoid the knowledge that tiger parts had been used for pharmacological applications for thousands of years. They might have considered adding a recommendation that tigers—Chinese and others—should not be recklessly slaughtered for medical purposes.

The *International Tiger Studbook*[*] reports that there were fifty-three South China tigers in captive facilities in 1999, but no figures are available for this species in the wild. In a report submitted to *Cat News* in 2003, the IUCN Cat Specialist Group wrote that "an eight-month field survey organized by China's State Forestry Administration in 2001–2002 failed to find any of this race of tiger in the wild. But it is thought that there might be some individuals surviving in the wild, although not forming a viable population with a long-term future." This suggests that the South China tiger, like the Bali tiger, the Caspian tiger, and the Javan tiger, is close to extinction, if it has not already arrived at that dark and irreversible destination.

As of 1997, the status of the South Chinese tiger in the wild could only be described as "vague," a term that was used in an article in *Oryx* entitled "The Decline and Possible Extinction of the South China Tiger" by Ronald Tilson and his colleagues. They wrote, "No wild South China tigers have been seen by officials for 25 years and one was last brought into captivity 27 years ago. . . . Over the last 40 years wild populations have declined from thousands to a scattered few." In 2002, an article in *Cat News* asked, "Is the South China tiger extinct in the wild?" and cited a survey of the Chinese Forestry Administration that declared there was some evidence of tigers in the mountains of South China. But Ronald Tilson, who participated in the survey, said, "By the end of the year, we had seen no tigers, not one. More surprising, we uncovered no recent physical evidence that provided incontrovertible proof of tigers. With only a few exceptions, we found little in the way of animals that tigers could eat." It would be impossible to say that every single South China tiger is gone, but when dedicated searchers spent seven months in eight of the most likely areas and failed to find even a pugmark, the message seems unequivocal: there are no more wild South China tigers.

---

[*] Figures from the *International Tiger Studbook* appeared in Vol. 32, p. 24 of *Cat News*, the newsletter of the Cat Specialist Group of the IUCN (the World Conservation Union), edited by Peter Jackson, Chairman of the IUCN/SSC Cat Specialist Group. Many of the articles are unsigned and are taken from news reports or studies published under other auspices. Attributions to articles that appeared in *Cat News* will be cited simply as "*Cat News*, Vol. #."

Killing tigers as pests and eliminating habitat certainly contributed to the downfall of this subspecies, but the Chinese tiger had the additional misfortune of living in the very country where its parts were among the most desirable components of traditional medicine. Not content to use their native tigers for local medications, China also became one of the largest *importers* of tiger products from Southeast Asian countries, and then turned around and exported the medicines: "CITES data for 1990–1992 show that China exported 27 million units of tiger products to 26 countries and territories" (Mills and Jackson 1994). The decline of wild tigers in China encouraged the establishment of "tiger farms" about which Tilson and his colleagues say, "It is reported that these tigers are being bred and maintained for eventual reintroduction efforts in northern China," but the high prices of tiger products—as contrasted with the nonremunerative nature of reintroducing tigers to the wild—causes one to wonder about the real purpose of these "farms."

One breeding center set up in 1986 at Hengdaohezi in northeast China was the subject of a 2000 report by Kristen Conrad. "The original objectives of the Center were twofold: preserve the species, while, in parallel, raise tigers for use in traditional Chinese medicine," she said. The tigers were ostensibly the "Siberian" variety (*Panthera tigris altaica*), which had been requisitioned from various Chinese zoos. By 1989, thirty-seven cubs had been born in the facility, of which twenty-eight survived. The plan was to release some into the wild to replenish the dwindling number of wild tigers in China. (In the zoos from which the tigers came, they might have been crossbred with other tiger subspecies, so these may not be pure "Siberian" tigers.) The second part of the plan was negated by China's compliance with the 1999 CITES treaty, which forbade the sale, purchase, transport, or pharmaceutical use of tiger products. With its main source of revenue thus lost, the center was unable to repay its loans and had to cut back on expenses. The breeding program was curtailed, and several tigers died of exposure and malnutrition.

As of a 1991 visit of E. B. Martin, L. X. Chen, and C. K. Lin (which they described in an article later that year in *International Zoo News*), the Hengdaohezi tiger center was already in financial straits, and "because [tiger] products have great value in Chinese medicine . . . the center would like to obtain permission from CITES to sell overseas some of the captive bred tiger products from animals which die of natural causes or disease." This plan seemed to the observers to be fraught with problems; there would be no way of identifying the products as having come from captive-bred tigers, and parts from wild-killed (poached) animals could be passed off as having come from those that died of "natural causes." CITES turned down the request. Under the terms of the Pelly Amendment, China was certified (charged for possible trade sanctions), and upon certification, China banned the sale of tiger (and rhino) parts. In 1996, wrote Conrad, the Chinese government argued that sustainable use to TCM is not incompatible with CITES regulations, but because allowing a legal market for tiger bones would have increased consumer demand, which in turn would encourage even more poaching of wild tigers, CITES again refused China's petition.

Somehow, the Hengdaohezi Center survived its financial problems, and, as of October 1999, Conrad continued in her report, "there were 147 live tigers associated with the Center, and an additional 46 preserved in storage." In 1996, 60 tigers were transferred to a new "Siberian Tiger Park" at Harbin, which (in 2004) its Web site declared, is "the largest wild natural park in the world for the northeast tiger; it is located on the northern bank of the Songhua River. A special tour car allows you to linger through groups of tigers and appreciate the adventurous and exciting scene." (At the 2004 IUCN World Conservation Congress, held in Bangkok, a resolution in support of Hengdaohezi was proposed, but it was withdrawn because of strong opposition from much of the conservation community.)

There are now a number of tiger farms/parks in China, and in December 2002, one hundred tigers (and two thousand crocodiles) were shipped from the Si Racha Tiger Zoo in Thailand* to the city of Sanya,

* According to the Thailand zoo's Web site: "At the Si Racha Tiger Zoo, 8 km east of town, you can hold a baby tiger in your arms, or have your picture taken with a tame crocodile. The Zoo covers more than 40 hectares and teems with wildlife, including more than 100,000 crocodiles and one of the world's largest groups of Bengal tigers, more than 130 of them. The Zoo maintains that it is the most successful program for breeding tigers in captivity anywhere in the world. Circus shows include pig racing, ostrich racing, and female performers who wrestle with crocodiles or cover themselves with scorpions. Try the crocodile satay or crocodile soup at the restaurant. Open daily from 9:00 am to 6:00 pm."

on China's Hainan Island, for a theme park known as "Love World." At a press conference held on December 26, 2002, the day after the tigers arrived in Sanya, a spokesperson for Sanya Maidi Ltd., a Sino-Thai joint venture, was quoted as saying, "After we have bred tigers for a few years, we might have over 1,000 of them. Tourists are likely to eat tiger meat at Sanya." This announcement caused such an uproar in the press that theme park officials quickly denied that such a statement had ever been made and claimed that the tigers were being held only for research and the edification of the public. Chinese State Forestry Administration officials promised to look into the matter, and Sanya Maidi threatened to sue the media for publishing the "fabricated story."

As of late 2003, the designation of worst abuser of endangered species seems to have shifted from Taiwan to Thailand. In a report published in the *Washington Post* for December 10, Ellen Nakashima described raids on private homes and zoos during the previous fall that revealed Thailand as "a major gateway for a thriving international trade in endangered species." More than *thirty-three thousand animals*, including tigers, bears, orangutans, and birds, have been recovered, and animal smuggling has been shown to be second only to drug trafficking in profitability. In October 2003, a house was found crammed with tiger carcasses quartered and on ice and six more starving tigers in cages, twenty-one severed bear paws ready for the pot, five live bears, and many other kinds of starving mammals and birds, all evidently destined for restaurants in Bangkok, where tourists are promised exotic dishes. Among the carcasses was a saola, a newly discovered deerlike animal, and a baby orangutan in a freezer. In a raid on Safari World, a park in southern Thailand, police found 144 orangutans, although the zoo had registered only 44. At the Si Racha Tiger Farm, officials found several hundred animals that the owners could not account for. Thailand's Queen Sirikit and Prime Minister Thaksin Shinawatra said that they deplored the illegal wildlife trade but did not offer suggestions on how to patrol the country's 1,800 miles of porous borders with Myanmar, Cambodia, Laos, Vietnam, Malaysia, and Indonesia.

Reported by the journal *Science* for October 29, 2004 (Anon. 2004a) and the Environmental News Service is the story of one hundred tigers that died of avian flu at the Si Racha Tiger Zoo, and the thirty more that were to be killed because they are showing symptoms. Dr. Charal Trinwuthipong, director of the Thai government's bird flu center, said that the tigers might have picked up the disease from infected raw chicken meat. Tigers and chickens, however, are not the only creatures to be affected by the strain of avian flu known as H5N1: from October 2003 to October 2004, this strain killed thirty-one people in several Asian countries, eleven of whom were in Thailand. The Si Racha Tiger Zoo was temporarily closed to the public on October 19, 2004. (Given the zoo's previous history, it has been suggested that there was no compelling reason to destroy thirty more tigers—unless their parts were destined to be sold as medicine.)

There is a substantial difference between "tiger farms" and tigers in captivity in Chinese zoos. The former are ostensibly being held for reintroduction into the wild but may also be used to supply illegal tiger parts, whereas the latter are for display and to save the subspecies from extinction. But even in zoos, the population of South China tigers is in decline, and because so many of the fifty captive tigers trace their ancestry to a single breeding pair, genetic diversity has been severely compromised. The wild population may already be functionally extinct, and the captive population, for better or worse, may be all that is left as a barrier against total extinction of this subspecies. Their extinction would certainly be hastened if their "parts" were made available for sale.

In *How the Tiger Lost Its Stripes*, a discussion of the politics of tiger conservation by journalist Cory Meacham, we read of Sam LaBudde (who posed as a cook on a Panamanian fishing boat in 1987 and videotaped the slaughter of dolphins in the tuna nets), who shot film in Taiwan that "showed flagrant violations of Taiwan's domestic wildlife trade laws, including shots of live tigers—which do not occur naturally on the island—pacing in cages at what LaBudde identified as a breeding farm." Subsequent evidence suggested that the animals were being butchered for sale, a practice that had been both legal and open in Taiwan until the 1980s but that was incontestably forbidden by the time LaBudde claimed the videos had been shot. A December 1997 article by Geoffrey Ward in *National Geographic*, a magazine known for its careful fact-checking, contained a double-page spread showing LaBudde holding up a photograph taken in Taiwan that showed a tiger in a cage "just before [it] was cut up for food and medicine, a practice that has declined there."

The Tigers of India

There are 827 million Hindus in India, more than 80 percent of the population, and the Hindu ethic is strongly respectful of animals. The attempt to protect tigers by establishing a network of preserves throughout the country has worked moderately well, but the runaway population explosion has overwhelmed the tigers. There are approximately five tigers for every 100 million people, and as the people's need for land and food increase, tigers are being crowded out of their ancestral homeland. As we've seen, it is not only the need for land that results in the displacement of tigers; it is also the need for tiger "products" that has made the tiger the prime target for poachers, and the great cats are being trapped, shot, poisoned, and electrocuted out of India—and out of existence. But before they disappear completely from the Indian subcontinent, the animals that Hindus call *bagh* have one last card to play: they bite back.

Throughout their range and throughout their history, tigers have been known to attack people and often eat them, which has resulted in the appellation "man-eaters." Nowhere has this problem been manifest more than in India, where overcrowding and competition for land use often bring tigers and humans into close proximity. In such an encounter it is the men (or women, or children) who suffer first, but the tiger usually loses in the end—except in India's Sundarbans region, where, as we'll see, man-eaters seem to be the rule rather than the exception and where retaliation is almost impossible.

From the earliest days of the British East India Company, the profusion of wild game in India amazed and delighted would-be hunters. British sportsmen killed deer of many species and wild boar, which they speared from horseback in the sport known as "pig-sticking." Tigers were the main attraction, however, because they (sometimes) attacked people and the hunt could therefore be justified as self-defense, but more often because they made wonderful trophies. The maharajas demonstrated their omnipotence by hunting tigers from the backs of elephants; beaters, sometimes numbering in the hundreds, drove the tiger into rifle range and the Indian prince shot it from the howdah. In *The Deer and the Tiger* (1967), George Schaller gave some of the more egregious examples:

> Gordon-Cumming (1872) shot 73 tigers in one district along the Narmada River in 1863 and 1864, and he once shot 10 tigers in 5 days along the Tapti River; Forsyth (1911) shot 21 tigers in 31 days in Uttar Pradesh; George V and his party shot 39 tigers in 11 days in Nepal in 1911–12; Rice (1857) shot or wounded 158 tigers, including 31 cubs, in Rajasthan between 1850 and 1854; the Maharaja of Nepal and his guests shot 433 tigers, as well as 53 Indian rhinoceros, between 1933 and 1940; Colonel Nightingale shot over 300 tigers in the former Hyderabad State; the Maharaja of Udaipur shot at least 1,000 tigers during his lifetime; the Maharajkumar of Vijayanagaram . . . has shot 323 tigers to date (letter, April 5, 1965); and the Maharaja of Surguja wrote to me in a letter dated April 6, 1965: "My total bag of Tigers is 1,150 (one thousand one hundred and fifty only.)"

During the long British occupation of India, sportsmen of one kind or another killed tigers for no reason other than that they were there. Of course, it was beneficial to the community to kill those tigers that were killing livestock, and even more accommodating to kill man-eaters, so the slaughter could be comfortably rationalized. Besides, the skin made a splendid trophy. As told in his 1946 *Man-Eaters of Kumaon*, the renowned Jim Corbett was invited in the first decades of the century to the Kumaon Hills, in the foothills of the Himalayas, to dispatch the man-eaters that were killing and eating the villagers. Corbett's first conquest was the "Champawat Man-Eater," a tigress that was said to have killed 434 people over a four-year period; then he got the "Chowgarh Tigers," a mother-daughter pair that had killed 64 people. He went on to kill many more, including the "Bachelor of Powalgarh," perhaps the largest Indian tiger ever measured, which does not appear to have killed anybody but because of his size, was "the most sought-after big-game tro-

The Prince of Wales on a tiger hunt in India, as drawn for the *Illustrated London News,* April 1, 1876.

phy in the province." In the 1930s, Corbett's publicized exploits encouraged Indian and European sportsmen to follow his example, and they killed so many tigers that the Indian government finally banned the killing of tigers for sport, but not before hundreds more had been shot. (Ironically, the 570-square mile Corbett National Park, located near the western border of Nepal, was established in 1936 as India's first national park, and in 1973, it would become the first reserve to be incorporated into Project Tiger.) Peter Hathaway Capstick, author of *Maneaters* (1981), believes that killing tigers was necessary to protect people from the vicious beasts of his eponymous title:

> Between the Brits and the Maharajahs there's pretty good reason to believe that at least 100,000 tigers got the deep six over the past hundred years. (Personally, since this is only 1,000 per year, I can't see the harm.) Some researchers will tell you that this number was killed since the beginning of the 20th century alone, and when we look at the number of survivors through the haze of now-settling dust of the

> great Terai [an area that formerly extended over the foothills of India and Nepal], they may be right. Through the incredibly expensive machine that was the Indian tiger hunting establishment, official guesstimates (now a couple of years old) are that, in India, of the some 40,000 left around 1930, less than 2,000 may yet exist.

In India, the human population is now well over a billion, and despite a vast migration into the cities, millions still live in or near undeveloped forests or grasslands, where there are wild animals to prey on them. And it has always been so in India: "In the past four centuries," wrote Peter Matthiessen, "tigers are thought to have killed 1,000,000 Asians, or about 2,500 people annually, or twenty-five people per 1,000 tigers—not an unreasonable figure when one considers that a man-eater of yore would often kill that many and more all by itself. In the past century, of course, human mortalities have declined for want of tigers, despite the increase in density of this form of prey."

The tiger's inclination to attack and sometimes eat people has been shown to be an impediment to its protection in India and Nepal, and throughout its range, though the Amur tigers of the Russian Far East do not seem to have much of a history of man-eating. In *Conservation Biology* (1997), Vansant Saberwal pointed out that "problem tigers that become habitual man-eaters . . . [are] one of the most basic causes of local animosity toward tiger conservation." People might be more inclined to save animals if the animals didn't sneak up on them in the dead of night and kill them. In 1998, Ronald Tilson and Philip Nyhus reported on the situation in Sumatra, where "in 3 months, four villagers were killed and five villagers were attacked by tigers in one multiple-use protected forest. In 1996 and 1997 more than a dozen deaths were allegedly caused by tigers in Sumatra, far above the average of two per year cited by Indonesian authorities."

Neither big game hunting nor self-defense can account for the recent decline in tiger numbers throughout India and Southeast Asia, although they certainly contributed to it. The increased demand by TCM for tiger parts is now the main cause of the destruction of tigers, and there is a direct and unfortunate correlation between the increase of poaching and the decrease of tigers. This correlation can be seen even in those areas where the tigers are (supposed to be) protected by law.

The upsurge in poaching tigers for TCM can be attributed to several factors, which came together around 1985. Because Asians have been using tiger-bone medicines for thousands of years, it was not the promise of a new cure for disease that encouraged the poachers, but more likely it was the knowledge that tigers, hunted as trophy animals for centuries, seemed to be disappearing at an alarming rate. And as tigers became scarcer, the possibility of enormous profits increased. Those who would use TCM were increasing even faster than the tiger populations were decreasing: the human population of just China had risen to 1.2 billion. A resurgent interest in TCM and the concurrent decrease in the effectiveness of patrols of some Indian tiger preserves after the death of Indira Gandhi in 1984 combined to pose a terrible threat to the tigers of India. In her 2000 TRAFFIC report, Kristin Nowell characterizes the recent history:

> It is reasonable to say that, by the mid-1980s, China's bone bank was running low and, with a few Tigers left in the wild in China, demand for bones for the neighboring Tiger countries increased, perhaps explaining the poaching of tigers from Dudhwa [a preserve in India] in 1986. In 1992, Ranthambhore [another preserve] sprang into the headlines. Well-known tigers were no longer being seen. A member of a traditional hunting tribe was caught with a Tiger skin and skull. He disclosed that he took bones to a butcher in a nearby town, who was found to be in contact with illegal traders. Not long afterwards, in September 1993, investigations by TRAFFIC India led to seizures of caches amounting to 400 kg [880 pounds] of Tiger bones in the Tibetan quarter of Delhi, apparently bound for China across the Himalayan passes.

With the increase in poaching and the long history of killing tigers for sport or protection, how many tigers are actually left in India? According to India's Ministry of Environment and Forests' *Annual Report* for 2002–3, "In Project Tiger at present there are 27 tiger reserves covering an area of 37,761 sq. km, with a population of about 1,498 tigers. This amounts to almost 1.14 % of the total geographic area of the country." There may be another 2,000 tigers in various national parks and wild areas throughout India, for a grand total of 3,498. Because tigers are so difficult to count, however, the number in the parks and preserves (and the total for all of India) is really only an estimate—and a problematic one at that.

For the past thirty years, tiger population estimates were based on a method invented in 1966 by S. R. Choudhury. This required thousands of forestry department personnel to fan out across India annually, searching for tiger tracks and then making plaster casts of the tracks. These "pugmarks" were then compared and individuals were identified and counted. Led by K. Ullas Karanth, an international team of tiger

Antoine-Louis Barye was famous for his bronze animal sculptures, particularly of the big cats. Here a tiger attacks an elephant carrying tiger hunters. The fearlessness of the tiger was recognized even in parts of the world such as Paris, where tigers could only be seen in zoos.

researchers examined this methodology for counting tigers and concluded that using footprints (pugmarks) to identify individual tigers (and then add up the totals) was essentially useless.

The potential tiger habitat in India extends over more than 300,000 square kilometers (186,000 square miles), an area approximately the size of Spain, and it is difficult to imagine any group of researchers carefully examining an area of that size—especially since much of it is densely overgrown, tall grass, or rocks that hold no tracks. "Unless clear impressions of all four paws on the right substrate are detected *for each individual tiger*," wrote the authors, "it is impossible to pick the *same hind pugmark* of each individual for comparisons as prescribed by the pugmark method. In reality, census personnel often do not find clear prints of all four paws, and consequently lift prints of the different paws of the same animal from different localities." In other words, it is possible to identify pugmarks from the same animal as having come from different animals, inadvertently increasing the number of tigers.

Pugmarks cannot be used effectively to differentiate individual tigers either, it seems. In one Karanth study (1987), when researchers were shown tracings of tracks, some identified them as having been made by twenty-four different tigers. In fact, they were made by only four tigers. In an interview with Seema Singh in late 2003 for *New Scientist*, Karanth said the pugmark method of counting tigers was not scientifically defensible: "Although an elaborate record-keeping protocol has been prescribed for the pugmark censuses, this protocol essentially ignores the fundamental need for mapping and geo-referencing the tiger signs that are detected by field workers. As a result, even after 30 years of pugmark censuses, large scale, country-wide maps of tiger distribution are not yet available. . . . So three decades of tiger monitoring have basically failed in India, despite being backed by massive investments." When asked where all the money went, Karanth replied: "On corrupting the whole system." Did his paper have much impact? According to Karanth, "The Ministries of Forest and Environment have shown no interest," and the government's response has been "to do nothing."

The preserves established by the Indian government and Project Tiger were designed to provide a haven for tigers, a place where hunting would be prohibited and where human activity—with the possible exception of tiger watching—would be discouraged. Unfortunately, India's burgeoning population, deprived of room to expand, often invaded the preserves, and at the same time, usually for political reasons having little to do with tigers, guards and patrols were reduced. For many Indians, whose average annual income is around $400, an animal worth thousands of dollars on the TCM black market, passing within rifle range, would be all but impossible to pass up, regardless of the laws or guards. And the national publicity afforded the tiger preserves told the poachers exactly where to go. The situations differed from park to park, but many of them headed straight for Ranthambhore, probably India's best-known tiger preserve.

Once the private hunting preserve of the Maharajas of Jaipur and now one of the most famous of India's national parks, Ranthambhore includes some 250 square miles of low rolling hills and grasslands in southwestern Rajasthan. High on one of the hills is the thousand-year-old fortress that gives the reserve its name—a now-overgrown edifice that often provides a spectacular setting for the park's population of tigers. It has become the special province of tiger conservationist Valmik Thapar, who has already written two books about the tigers of Ranthambhore and produced a BBC special there, called "Land of the Tiger."

Ranthambhore was one of the jewels in the crown of Project Tiger; in the early 1970s, all the villagers inside the park's boundaries were moved out. In *The Wild Tigers of Ranthambhore*, Thapar wrote, "Tigers and human beings cannot share the same tract of forest. One of them had to go. Fortunately in Ranthambhore, the humans agreed to relocate." In 1974, there were only twelve to fifteen tigers in the park; by 1989 there were fifty. In the 1980s, the tigers of Ranthambhore became more visible—the nine feeding tigers and the interactions of Padmini and her mate, described earlier in the chapter, were observed in daylight in 1982; and the huge male known as Genghis was observed to have developed a unique method of hunting: he would charge right into the lake where sambar deer were resting and attack them in the water.

The tigers of Ranthambhore flourished briefly, but then the truce between tigers and men was broken in September 1988 when a young tiger killed and partially ate a seven-year-old boy on the fringes of the forest. And if this wasn't enough to rekindle antagonism toward the tigers, the poachers arrived. "Poaching, and the horrors of it," wrote Thapar in 1999, already "had percolated through the park and only in 1992, with the arrest of a poacher with a tiger skin, did I realize that at least 15–20 tigers had been poached over the last two years." The Spring 2003 issue of the IUCN's *Cat News* put the number of tigers surviving in Ranthambhore at thirty-two and described an additional threat to the remaining population: a significant increase in human visitors to the region, which increased the possibilities of poaching.

The Sariska reserve borders on the Jamwa Ramargh Wildlife Sanctuary in Rajasthan. According to a June 2003 report by Antony Barnett in the *Guardian Weekly*, Unilever and other firms have been mining talc (a magnesium silicate) in the sanctuary. Used extensively in the manufacture of soap, eye shadow, deodorant, body lotion, and baby powder, talc is produced by crushing giant soapstone boulders that have been extracted from the earth by the use of dynamite. Inhalation of the fine-powdered talc has proven to be a major health hazard to the workers, and also a major health hazard to the twenty-two tigers identified there in the spring 2003 issue of *Cat News*. By 2004, there were seventeen tigers remaining in the 300-square-mile reserve. A year later, there were none. According to a March 2005 Reuters report by David Friel, Indian Prime Minister Manmohan Singh convened a meeting of forest officials, wildlife experts, and community leaders to investigate reports that Sariska's tiger population had been completely wiped out by poachers. Four visits by the Committee for Forests and Wildlife failed to find a single tiger in Sariska.

Located due east of Ranthambhore and Sariska, Sundarbans is a unique and very special tiger habitat. Sundarbans covers some 10,000 square kilometers (6,200 square miles) of mangrove forest and water, 60 percent of which is in Bangladesh and the rest in India. The Ganges, Brahmaputra, and Meghna rivers converge on the Bengal Basin to create the world's largest delta. The entire Sundarbans area is intersected by an intricate network of interconnecting waterways of which the larger channels are often a mile or more in width and run in a north-south direction.

Rainfall is heavy and the humidity averages 80 percent because of the proximity of the Bay of Bengal. It is called Sundarbans because of the dominance of the tree *Heritiera fomes*, locally known as *sundari*. Sundarbans is the only remaining habitat in the lower Bengal Basin for a great variety of animal species. The region once boasted the Javan rhinoceros (*Rhinoceros sondaicus*) and the wild water buffalo (*Bubalus bubalis*), which were last recorded in 1870 and 1885, respectively, and are now considered extinct in the region. The swamp deer (*Cervus duvauceli*) existed in good numbers until early in the twentieth century, and the Indian muntjac (*Muntiacus muntjak*) was last reported on Halliday Island in the late 1970s. The tiger population in the Indian side, estimated at 270 in 1997, is the largest in that nation. The relatively high frequency of encounters with local people within the boundaries of the tiger reserve is probably largely responsible for the notorious man-eating reputation of the Sundarbans tiger. In an unusually poetic style for a scientific presentation, John Seidensticker (1987b) wrote of Sundarbans:

> When the water drains off the higher ground with the ebb and steel-gray mudbanks are exposed and shimmering under a scorching mid-day sun, the place to look for a tiger is in the shade at the mouth of small, side khals. And if you are truly fortunate you see it: head raised, lying half-submerged, intently watching as you slip by in a launch—a classic Sundarbans tiger.

The tigers of this region are the most dangerous in India. Sy Montgomery, who journeyed to that great delta and came back to New Hampshire to write *Spell of the Tiger: The Man-Eaters of Sundarbans*, gives a vivid description of these animals:

> And here, the tigers do not obey the same rules by which tigers elsewhere govern their lives. They hunt people. They take their prey often in broad daylight. They will even swim out into the Bay of Bengal, where the waves may be more than two feet high. They often swim from India to Bangladesh. The tigers here are bound by neither day nor night, land not water; these tigers, some say, are creatures of neither heaven nor earth.

Their proclivity to enter and remain in the water distinguishes Sundarbans tigers from all others. While other tigers scent-mark their territories by spraying or scratching trees along well-worn paths, the marsh tigers' territory is largely under water, so how they mark their territory—or if they have defined territories at all—is unknown. And where other Indian tigers prey on deer, wild pigs, lizards, and birds, the tigers of Sundarbans seem to have a penchant for eating people. "In the old days," wrote Richard Perry in 1965, "it was the salt boilers, working on long sand-spits projecting from the jungles, who were the chief victims, despite the look-out posted to watch for tigers; and one tigress, who was determined enough to leap an eight-foot stockade for her victims, seized sixty men in the space of three months." Currently, the "official" number of attacks in Sundarbans is thought to be around thirty or forty per year, but, as Montgomery points out, some unofficial totals for the whole Sundarbans region run as high as 300 per year. She writes:

> No one except the Forest Department officials are allowed inside Sundarbans Tiger Reserve's 514-square-mile core area, which is set aside for wildlife alone. Ringing the core is a buffer zone of 562 square miles, where people may fish, collect honey, and cut wood, but they must have a permit to do so. If a permit holder is killed inside the buffer zone, the government compensates the family for the loss, and the death is officially tallied. But families of tiger victims who are illegally inside restricted areas receive no compensation, so there is no reason for them to inform the authorities; in fact, these families fear they might be prosecuted.

As for *why* Sundarbans tigers are more likely to attack people than tigers elsewhere in India, nobody really knows. In his 1987 discussion of man-eaters, Charles McDougal cited a 1975 study by Hubert Hendrichs that suggested drinking salt water—there is hardly any fresh water in Sundarbans—might make the tigers more ferocious and more likely to attack humans. Others have suggested that because the tigers often eat fish from the nets of fishermen, perhaps they might associate people with food, or that in the difficult hunting conditions of the region, humans were an especially inviting and available target. Peter Matthiessen, in *Tigers in the Snow*, gives us something else to consider as well:

> A paradox still poorly understood is not why tigers attack human beings but why they do not attack more often, all the more so now that the senses and agility of *Homo sapiens*—always rudimentary when compared to those of other mammals—have been further dulled and softened by modern life. No matter how athletic and alert, a man or a woman inevitably presents a large, slow, easy prey to a hunting tiger.

Although there are no settlements within the boundaries of Sundarbans, people still risk their lives to cut wood, catch fish, or gather honey there. Protection from killer tigers often takes the form of prayers, but in addition, pigs have been released to keep the tigers well fed, freshwater ditches have been dug for the tigers to drink from, and dummies that are wired to shock an attacking tiger (the fuse immediately blows so that the cat will not be electrocuted) have been deployed. Face masks have also been provided for people to wear on the backs of their heads while working in the forest, because it is believed that tigers will only attack people from the rear. These worked, wrote Peter Jackson (1990), "as long as the tigers believed in them." He quotes a researcher who said, "After five or six months, they were finding out that this was not the front of the human being. . . . They know what a human being looks like. They know here is a back and a front. They are finding out that this is not a good front." Montgomery comments: "Nothing—neither laws nor permits nor patrols—stops men from illegally crossing into the reserve's forest core; and nothing—neither offerings nor armor nor trickery—stops the tigers who come to meet them."

The estimated 270 tigers in the Indian Sundarbans cover only the 40 percent of the delta that is in India; the remainder is in Bangladesh, where there are no current figures for tigers. Estimating tigers anywhere is fraught with difficulty as we've seen, but in Sundarbans it is made even more complex by the difficulty of tracking tiger movements through tidal waters. Even if the pugmark method of counting tigers worked, it would be useless in Sundarbans, where the tigers swim from place to place, and even if they were to walk along the mudbanks, their tracks would quickly be obliterated by the surging tides. On July 31, 2003, on the BBC News Web site (http://newsvote.bbc.co.uk) there appeared an announcement of a joint tiger census to be taken in the more than 10,000 square kilometers that make up the Indian and Bangladeshi Sundarbans, but how they were actually going to count the tigers was not explained.

Kaziranga, in the northeastern state of Assam, is a national park and therefore a de facto tiger reserve. It is also the home of the world's largest remaining population of one-horned rhinos, as well as wild buffaloes, wild boars, elephants, swamp and hog deer, leopards, hoolock gibbons, capped langurs, rock pythons, monitor lizards, various species of eagles, partridge, storks, herons, and assorted waterfowl. The riverine habitat consists primarily of tall, dense grasslands interspersed with open forests, interconnecting streams, and numerous small lakes or *bheels*. Three-quarters or more of the area is submerged annually by the flood waters of the Brahmaputra. In Kaziranga's 266 square miles there may be as many as eighty-five tigers. Without a buffer zone, and with many impoverished people on its border, the temptation to poach tigers in the park is particularly great.

The overall situation for tigers in India has become critical, in no small part because of habitat loss. In an essay in the 1999 compendium *Riding the Tiger: Tiger Conservation in Human-Dominated Landscapes*, Valmik Thapar warns:

> Fragmented and deeply scarred, this is the land of the tiger and is home to 50% of the world's wild tiger population. Comparing past and present forest maps of India, one can see that large chunks of forest have now ceased to exist in the west and north. We are ending up with little islands that may or may not survive the pressures of time. Whether it is the great floodplains of Kaziranga, the monsoon and rainforests of peninsular India, the mangrove swamps of the Sundarbans, the lower Himalaya in Manas or Corbett, the problems of these areas mount and we enter an area where the tragedy of the tiger overwhelms. It is time to start from scratch.

His conservation suggestions involve "emergency action, reform and a total change in the system of wildlife governance" and require identifying *fifty* tiger sites in India, completely revising the administration of the protected areas, training forest guards, creating educational programs and community-based activities to encourage local people to protect their natural heritage, and careful monitoring of all these activities. In 2003, Thapar concisely described the shifting statistics for tigers, reserves, and humans in India: "The year 2003 is the 30th anniversary of Project Tiger. When Project Tiger started in 1973 the population of India was about 780 million—today it is nearly 1.1 billion. There were nine tiger reserves in 1973 and 29 in 2002. There were an estimated 1,800 wild tigers. I believe that today, 30 years later, there are about the same number alive—maybe a few hundred more." As increased poaching reduces the number of tigers, the total population will be gradually reduced until the big cats approach extinction. True enough, wrote John Kenney and his colleagues in 1995, but the decline may not be gradual. As the numbers continue to fall, "the probability of population extinction increases [because] a critical zone exists in which a small, incremental increase in poaching greatly increases the probability of extinction. The implication is that poaching may not at first be a threat but could suddenly become one." Using sophisticated modeling techniques ("individually based stochastic spatial simulations"), the authors examined a numbers of variables, including large and small population sizes, known breeding rates, long-term and short-term poaching, and the cessation of poaching altogether, and found—not particularly surprisingly—that "poaching reduces genetic variability, which could further reduce population viability due to inbreeding depression, [and] the longer poaching is allowed to continue, the more vulnerable a population will be."

Nagarahole was where I saw my first wild tiger in 1997. In May 2003, it was announced that the government of the state of Karnataka intends to abolish the post of field director of the Bandipur-Nagarahole Tiger Reserve and place the reserve under the control of the adjoining territorial administrators. This means that the reserve will virtually cease to exist and the forest will be managed by officers from the territorial department, instead of having a dedicated field director whose main responsibility is the health of the park. Project Tiger has warned the Karnataka government that such a move would mean that the park would lose out on financial and technical expertise from the center.

Originally established in 1931 as a sanctuary of 90 square kilometers, the land became Bandipur National Park in 1973, one of the original tiger reserves. The following year, Nagarahole was enlarged and upgraded to the status of national park, and as many as one thousand squatters were voluntarily resettled. Bandipur-Nagarahole is also part of the Nilgiri Biosphere Reserve, the first biosphere reserve in India. According to the Indian conservation magazine *Sanctuary*, the change in administration "will have the disastrous effect of virtually closing down the Bandipur Tiger Reserve, one of India's oldest and finest Protected Areas. . . . The park is home to a population of more than 80 tigers and is part of a globally-significant tiger belt that includes the Nagarahole National Park in Karnataka and the Mudumalai forest in Tamil Nadu."

The recent downgrade of Nagarahole suggests that the government of India, rather than encouraging tiger protection, seems prepared to discourage it. Revising the status of this preserve signals a new era in India's conservation politics—one that bodes ill for its tigers. Poachers killed eighty-one tigers across India in 1999, fifty-three in 2000, seventy-two in 2001, and forty-three in 2002. If the countrywide population estimates are wrong because the methodology of counting was inadequate, it appears that the only available accurate figures are for the number of tigers killed every year—and these numbers might be wrong too. In the Spring 2001 issue of *Cat News*, Valmik Thapar said in response to news of annual deaths from poaching: "The figures are terrible, but the ground realities are worse because a number of deaths still remain unrecorded." He added as a benchmark that the country had about 3,000 tigers at the start of the year 2000. Project Tiger Director P. K. Sen said that tiger numbers were declining because of "ever-growing habitat vandalism, depletion of the tiger's prey base, illicit trade in tiger parts, lack of infrastructure facilities, staff and money to effectively protect the tiger."

All the world's wild tigers are in considerable peril. In 1986, when several tigers were poached in the northern Indian preserve of Dudhwa, there was speculation that they might have been killed for their bones. In the same year, it was reported in Chinese newspapers that the authorities were establishing a tiger farm to provide bones for medicines. When the Chinese Communist Party gained power in 1949, the South

China tiger was seen as a threat to agricultural development. It was declared a pest and teams were appointed to hunt it down. According to Chinese scientists, more than three thousand skins were handed in during the 1950s and 1960s. A few tigers remained and were given legal protection, but the population had been virtually exterminated. Although China had a massive stock of bones, by the mid-1980s, the bone bank was running low, and, with few tigers left in the wild in China, the demand for bones from neighboring tiger countries increased. In 1992, Ranthambhore sprang into the headlines when well-known tigers were no longer being seen. In September 1993, investigations by TRAFFIC India led to seizures of caches amounting to 400 kilograms of tiger bones in the Tibetan quarter of Delhi, apparently bound for China across the Himalayan passes. Meanwhile, the Soviet Union had collapsed and with it law and order in the Russian Far East. In this remote outpost, the remaining tigers, which were few and far between, were hunted for their bones, to be sold to commercial dealers who would in turn sell them to "pharmacists." Investigation of customs records in South Korea revealed that hundreds of kilograms of tiger bones had been legally imported in the years leading up to 1993, all in the name of Chinese medicine.

The Tiger in Medicine

The demand for medicines delineated in ancient Chinese traditions is the engine that drives today's insatiable commercial market for tiger bones and other tiger parts, but the Chinese are not alone in using parts of tigers as medicine. In the *Indian Materia Medica*, which includes Ayurvedic, Unani, and Indian home remedies, tiger fat is listed as a treatment for leprosy and rheumatism; tiger claws can be used as a sedative, tiger teeth for fever, and tiger nose leather for dog bites. Tiger bone is used in Vietnam to make a balm that is said to help assorted ailments, including rheumatism and general weakness.

Here is the basic description of the tiger in Bernard Read's (1931) translation of Li Shih-chen's sixteenth-century materia medica *Pen Ts'ao Kang Mu*:

> It is the king of the mountain animals. It is shaped like a cat and is the size of a cow. Sharp bristle-like whiskers. The tongue is large and broad and full of spikes. Short-necked and squat-nosed. At night one eye is phosphorescent and provides light while the other eye is used for observation. It roars like thunder and causes the wind to rise. It enters the rutting season in the first week of November, others say it comes when the moon is cloudy and that they only copulate once in a lifetime. Gestation is seven months. The tiger has the power of divination and can sense direction, mark the ground and find food thereby. Men have learnt their ways. In their diet they follow the lunar calendar. During the first half of the month they eat the head end of an animal, during the latter half of the month they eat the tail end. . . . Pai Pu Tzu said that after 50 years tigers turn white. Old stories often tell of tigers turning into men and vice versa.

Tiger parts were prescribed for everything, but the most potent parts were the bones. Li lists every part that could possibly be utilized and specifies its applications:

> TIGER BONES. The yellow ones from the males are best. Animals shot with arrows should not be used because poison enters the bones and blood and is harmful to people. . . . The bones should be broken open and the marrow removed. Butter or urine or vinegar is applied, according to the type of prescription, and they are browned over a charcoal fire. Acrid, slightly warming, nonpoisonous. For removing all kinds of evil influences and calming fright. For curing bad ulcers and rat-bit sores. For rheumatic pain the joints and muscles, and muscle cramps. For abdominal pain, typhoid fever, malaria, and hydrophobia. Placed on the roof it can keep devils away and so cure nightmares. A bath in tiger bone broth is good for rheumatic swellings of the bones and joints. The shin bones are excellent for treating painful swollen feet. It is applied with vinegar to the knees. New born children should be bathed in it to prevent infection, convulsions, devil possession, scabies and boils, it will then grow up without any sickness. It strengthens the bones, cures chronic dysentery, prolapse of the anus, and is taken to dislodge bones which have

become stuck in the gullet. The powdered bone is applied to burns and to eruptions under the toenail.

TIGER FLESH. Said to be bad for the teeth. It should not be eaten in the first lunar month. For nausea, improves the vitality, and stops excessive salivation. For malaria. A talisman against 36 kinds of diseases. A tonic to the stomach and spleen.

TIGER FAT. For all kinds of vomiting. For dog bite wounds. Applied in the rectum for bleeding hemorrhoids. Melted and applied to scabby and bald-headed conditions in children.

TIGER BLOOD. It builds up the constitution and strengthens the willpower.

BILE OF THE TIGER. For convulsions in children.

EYEBALL OF THE TIGER. Eyeballs are not used from animals that have died of sickness. The ball is macerated overnight in fresh sheep's blood, and then separated and dried over a low flame, and powdered. For epilepsy, malaria, fevers in children, and convulsions. For quieting nervous children. It clarifies the vision and removes membranes over the eye. It stops crying.

THE NOSE OF THE TIGER. For epilepsy and convulsions in children. Hung on the roof it will induce the birth of boys.

TIGER TEETH. Applied to sores on the penis and running sores. Taken for hydrophobia and phthisis [wasting disease or tuberculosis].

TIGER CLAWS. With the bones and hair of the paw of the male animal tied on a baby's arm as a talisman.

TIGER'S WHISKERS. Given for toothache.

Bensky and Gamble's 1993 *Chinese Herbal Medicine* lists only tiger bone (*Os Tigris*) in the category of "Herbs That Dispel Wind-Dampness" and say that tiger bone can be used for "stiffness and migratory pain in the joints. [It] disperses wind-cold and strengthens the sinews and bones: for paralysis, weak knees and legs, spasms, stiffness and pain in the lower back, and cold pain in the bones." There is no mention of impotence. In *Of Tigers and Men*, Richard Ives gave this summary of the use of tigers in Chinese medicine:

> Since 1990, nearly half the surviving tigers in Siberia have been slaughtered, poached for their body parts, and sold to Chinese merchants who turn virtually every square inch of the tiger carcass into "medicines" that can be bought over the counter in Chinatowns in cities the world over. A compelling fund of evidence now suggests that most of the tigers killed in the wilds of Asia eventually end up in Chinese hands. Though it may seem like science fiction, the Chinese lust for tiger parts is so acute that with the extinction of wild tigers now in view, farms designed to raise tigers for slaughter are already operating in Taiwan. The discovery of such farms on the Chinese mainland itself would come as a surprise to no one.

There are certainly tiger farms in northern China, as we've seen, ostensibly to ensure the survival of the Chinese subspecies, and there may be (or may have been) tiger farms in Taiwan for less-commendable reasons. "Public slaughters in the mid-eighties were common throughout Taiwan," wrote Keith Highley in 1993. "In a four-month period in 1984, seven tigers were slaughtered, their parts auctioned off to the public. The slaughters were well-advertised in advance: ads were taken in local newspapers, men paraded through the village banging drums, and on one occasion a tiger slated for slaughter was paraded through villages in the bed of a pick-up truck." The most desirable of tiger parts is the penis, soaked in brandy for six months and then served as a tonic guaranteed to enhance a man's sexual prowess. Just as Taiwan was the center of the trade in rhino horn, as we saw in the discussion of rhino medicine, a decade ago, it was also the epicenter of the Asian trade in tiger bones. In 1993, Keith Highley visited 115 shops on Taipei's Di Hua Street and

These tiger-bone and sea horse pills were made in Japan, showing the appeal of TCM medications in other parts of Asia.

found that "68 shops (59.13%) were either in possession of tiger bone or offered to purchase it for investigators." The following year Keith and Suzie Chang Highley produced a further report:

> In the last decade, newly wealthy Taiwan has emerged as a leading consumer of endangered species products. By 1989 the country had

> been dubbed the "greatest threat to the survival of Africa's rhino." Taiwanese pirate whaling vessels killed whales and shipped substantial quantities of meat to Japan, laundering the contraband by way of transshipments through the Philippines and South Korea. A turnaround industry for ivory thrived; substantial amounts were imported (legally and illegally), worked in-country, and sold domestically or re-exported. Taiwanese drift net vessels frequently violated other nations' exclusive economic zones to poach salmon and other fish and, on occasion, capture penguins, seals, and other wildlife. Even the giant tridacnid clam was not exempt from the feeding frenzy; it was reportedly pushed to near extinction in Micronesia and other parts of the Pacific by Taiwanese poachers.

In Cambodia, a desperately poor country, the remaining tigers are being hunted almost out of existence mainly for sale of animal parts. Tigers shot by Cambodian and Vietnamese soldiers find their way to the marketplace at Phnom Penh. In December 1993, an entire carcass was sold in the market for US$1,500, and the buyer then sold the claws, skin, and bones separately for a handsome profit. An article in the Spring 1999 issue of *Cat News* tells of the use in northeastern Cambodia of homemade land mines to kill tigers for sale of their parts on the black market. Along a trail with recent tiger tracks, a fence is built with two entrances and a dead sambar deer placed between the entrances. At either end, a trip wire is attached to a mine, to be tripped by the tiger as it approaches the bait. "The threat of damaging the pelt—worth US$100 or more to the poachers—does not worry them. It is the bones that matter. Vietnamese traders across the border sell tiger bones for up to US$350 a kilogram, for consumption in Vietnam, China, and other international markets."

Of conservation efforts in Cambodia, Antony Lynam (2004) wrote that hunting is the single greatest threat. During the civil conflict of the 1970s and 1980s, there was a massive influx of weapons, and guns were easy for rural Cambodians to acquire. Populations of tigers, elephants, deer, and wild cattle were decimated for local consumption and the wildlife markets in Vietnam and China. Tracts are still littered with wire snares and traps set for killing animals, and illegal loggers, who hunt as they fell timber, are as dangerous to wildlife as poachers.

Woman selling tiger bones, Vietnam, 1994.

The Leopards

As the tiger populations are reduced, poachers and suppliers of material for TCM are forced to look elsewhere for the animal parts they require for their business. In the forests of southern Asia and Africa, and also in the snow-clad mountain ranges that form the spine of central Asia, there are other large cats whose spotted coats are even more desirable than the tiger's stripes and whose bones are now being incorporated into the TCM pharmacopoeia. Leopard skins have always been desirable commodities because of their spectacular spotted patterns. A recent survey published in the IUCN's *Cat News* (Spring 2003) gives the quotas for leopard skins legally exported from various African countries: Tanzania, Ethiopia, and Zimbabwe each have quotas of 500, while the total for all countries in Africa in 2002 was 2,335. Skinning 2,335 leopards leaves a lot of meat and bones behind, and it is not surprising to learn that the bones have become valuable trade items as well. Leopards (*Panthera pardus*) are being killed legally and illegally throughout their range, but the decline of tigers throughout Asia has recently encouraged the hunting of leopards specifically for their parts.

Curiously, there are very few uses listed for leopard parts described in manuals of traditional Chinese medicine: eating the flesh "keeps away evil diseases and benefits the kidneys"; the fat can be used to promote the growth of hair; if you boil the leopard's nose with that of a fox, you will expel fox-devils; the cranium can be used as a pillow to keep away evil spirits; but the skin may not be used as a sleeping mat because it "causes fear to enter one's soul, and the hair if it enters open wounds is poisonous, but it has been known to keep away ghosts" (Read's translation of *Pen Ts'ao Kang Mu*). In Bensky and Gamble's *Chinese Herbal Medicine* (1993), we read: "The bone of the leopard, *Os Leopardis*, can be used as a substitute for *Os Tigris*. It is acrid and warm and strengthens the sinews and bones, expels wind-dampness, and alleviates pain. It is not as potent as *Os tigris*. Dog bone, *Os Canis*, is also sometimes used, but it is considered excessively hot."

Evidence for the illegal leopard trade has appeared from Calcutta to London. On May 12, 2002, in London's *Observer* it was reported that an illegal trade in leopard parts was uncovered by police. They found arthritis medicines derived from leopard bones and twenty-five packets of medicated plasters made of ground leopard bones in an Asian supermarket in Hackney, East London, which is becoming a notorious center for the illegal international trade in endangered animals. And in India, on May 30, 2003, forest officials, assisted by the Wildlife Protection Society of India, seized two tiger skeletons, a leopard skeleton, and a leopard skin, in a series of raids carried out in Maharashtra in central India. Six kilograms of tiger bone were seized in one village, and one leopard skin, 6 kilograms of leopard bones, and 8 kilograms of tiger bone were found in two others.

At a maximum weight of about 150 pounds, leopards (and their bones) are considerably smaller than tigers, and now that leopards are being poached, they are becoming rare everywhere in Asia except perhaps India. These beautifully spotted cats occupy a much broader range than tigers; they are found throughout central and southern Africa and in Asia from Arabia to India, Southeast Asia, and China. As with the tiger, there are several recognized subspecies, and although no black tigers have ever been recorded, melanistic leopards (sometimes known as "black panthers") are not uncommon, particularly in Asian forest habitats.

The snow leopard is a large, ghostly gray cat that lives in the high mountain ranges of central Asia from northwestern China to Tibet and the Himalayas. It has been placed in a separate genus (*Uncia*) from the lion, tiger, leopard, and jaguar, all of which are classified as *Panthera*.* The largest population—perhaps two thousand—is found in China (mostly in the Tibetan region), but snow leopards are also found in the mountains of Afghanistan, Bhutan, India, Kazakhstan, Kyrgyzstan, Mongolia, Nepal, Russia, Tajikstan, and Uzbekistan. The world population of snow leopards has been estimated at somewhere between 4,500

* On the IUCN's Red List of Endangered Species, only one cat—the Iberian lynx—is "critically endangered," but four other felids are "endangered," where the remaining number is less than 2,500 and the population is "declining and fragmented." They are the Andean mountain cat, the Borneo bay cat, the snow leopard, and the tiger.

Women of fashion have always been partial to coats made from the skins of spotted cats. As she prepared to depart for a trip to India and Pakistan in 1962, Jackie Kennedy wore this leopard-skin coat.

and 7,000 animals, but in their high-mountain fastness, they are extraordinarily difficult to find, let alone count. The evidence suggests that China is also the biggest market for the snow leopard bones that are used in TCM, for skins, and probably live specimens. Even the consumption and trade of snow leopard meat has been reported. Pelts, bones, and live animals on the Chinese market originate not only from China, though, but also from the neighboring central Asian countries of Nepal, India, and Pakistan (Dexel 2002).

Trade in snow leopards extended to Kyrgyzstan and Tajikistan as well, at least after the breakup of the Soviet Union in 1991, and inter-

Snow Leopard (*Uncia uncia*).

est in these thick-furred ghost cats was anything but academic. Evgeniy Koshkarev and Vitaly Vyrypaev (2001) explain:

> Snow leopards were killed even in zoos, announcements about the sale of pelts and live animals were published in newspapers, and at the Kyrgyz Academy of Sciences a story circulated that a pregnant female snow leopard had been offered for sale by telephone. It was corruption and unemployment that turned the country into one huge black market, and made poaching and mediation in the sale of goods the only source of income for many of the inhabitants, urban as well as rural. There is no other way to explain the appearance of 200 snow leopard pelts in the Manas holidays of 1995, the massive sale to the Chinese market of medical raw materials made from the snow leopard in the 1990s and the absence in 1999 of snow leopards from the best habitat areas of Kyrgyzstan and Kazakhstan.

When Koshkarev and Vyrypaev conducted a survey of the five best-known snow leopard habitats in Kyrgyzstan, they found fresh tracks in only one. "If the situation in other Central Asia republics is close to that observed in Kyrgyzstan," they concluded, "then we are talking about the destruction of no less than half of the population."

All may not be lost in Kyrgyzstan, however, as a 2003 *Wildlife Conservation* article by Christopher Pala points out. The German conservation organization Naturschutzbund (NABU) funds the snow leopard conservation group known as *Gruppa Bars*, which is trying to curtail snow leopard poaching by turning poachers over to the local police and confiscating pelts that would otherwise be for sale. Under Birga Drexel, NABU's snow leopard project director in Germany, Gruppa Bars has opened a snow leopard compound at Lake Issyk Kul, where live leopards, illegally captured for zoos, can be rehabilitated.

The snow leopard's prey includes wild sheep, wild boar, hares, deer, marmots, mice, and other small mammals. There seems to be no record of snow leopards attacking people, but these cats also prey on domestic livestock, and the retaliation of herders endangers them. But nothing threatens the snow leopard more than hunters who kill them for their luxurious coats. Black market pelts are easily found in central Asian bazaars, and a full-length coat, consisting of six to ten full skins, can command as much as $60,000. The luxuriously thick, smoky gray coat of the snow leopard is quite different from the short-haired coat of the "ordinary" leopard, but with the exception of the skull, which is quite different, the bones of both kinds of leopards are remarkably similar, so the bones of snow leopards are now in demand as substitutes for tiger bone in Chinese medicine. Traders will pay up to $190 for a snow leopard skeleton in Tibet, and in northern Nepal—where the daily wage for sherpas is about $10—people have been seen to trade snow leopard bones for sheep along the border with Tibet.

A TRAFFIC report entitled "Fading Footprints: The Killing and Trade of Snow Leopards" (Theile 2003) reviews the history of these high-mountain cats and their new popularity in the TCM pharmacopoeia:

> Leopard bones, including Snow Leopard bones, have been used in traditional Asian medicines for centuries for a variety of treatments, including rheumatism, tendonitis and bone fractures. They are con-

> sidered to have acrid and warm properties and to have anti-inflammatory and pain-relieving effects. In the Chinese materia medica they are referred to as *Bao Gu* or *Os leopardi* and their properties are distinguished from those of Tiger bones, although they can be used as a substitute for the latter. The skulls of Snow Leopards have also been used in ritual ceremonies in parts of China and in Nepal and body parts other than bones, including the sexual organs, teeth, claws and meat appear in trade, for medicine and shamanistic practices.

Even the lonely mountains of the Himalayas offer no sanctuary for the snow leopards whose bones are eagerly sought for Chinese medicine. Poaching for their coats and bones has greatly reduced the populations of the great spotted cats of Africa and Asia. The very existence of the world's five rhino species is threatened by the insatiable demand for their nose-horns, some for dagger handles, but now mostly to be ground into medicinal powders that are believed to cure a variety of ailments. Where once there were eight subspecies of tigers in Asia, there are now five, and three of these are close to extinction. They are being hunted because traditional Chinese medicine needs the bones, flesh, fat, eyeballs, teeth, and claws, and fashion needs the pelts. Chinese medicine also decrees a need for selected products from certain bears, but unlike rhinos and tigers, the bears do not have to be killed—at least not immediately. Instead, they are put into such dreadful situations that shooting them would be a blessing.

6

The Bad News for Bears

Along with all species of rhinos and tigers, the world's wild bears are threatened by traditional Chinese medicine's demand for parts of animals. There are eight species of bears: the polar bear (*Ursus maritimus*), brown bear (*Ursus arctos*), black bear (*Ursus americanus*), spectacled bear (*Tremarctos ornatus*), sun bear (*Helarctos malayanus*), sloth bear (*Melursus ursinus*), Asiatic black bear (*Ursus thibetanus*), and giant panda (*Ailuropoda melanoleuca*). (For many years, the giant panda was classified with the raccoons, but comparative DNA analysis shows that it is indeed a true bear, and therefore its common appellation, "Panda Bear," is actually correct.) With the exception of the panda, all bear species are in danger in the wild because of the growing demand for bear gall bladder and bile, but none more so than the species that have the misfortune to reside in the very countries where their parts are so highly valued.

Sloth bears are found throughout India and Sri Lanka, and in Bangladesh, Nepal, and Bhutan. In northern India and Nepal they are called *bhalu*, undoubtedly the source of the name "Baloo," the bear in Kipling's *Jungle Book*. They have a long shaggy coat, usually black, but sometimes with enough brown or gray hairs to give them a tawny or grizzled appearance. They usually have thick ruff around the neck and a light-colored U- or Y-shaped patch on the chest. The muzzle is lighter in color. Sloth bears can be 6 feet in length, stand about 3 feet high at the shoulder, and weigh up to 300 pounds. They feed extensively on termites, and to accomplish this, they are able to protrude their naked lips, form them into a tube, and then suck in the termites through the gap made by a missing pair of upper incisors. Sloth bears also eat other insects, eggs, honeycombs, carrion, and various kinds of vegetation. In Nepal, they eat fruits extensively when in season, from March to June.

Throughout India, an itinerant people known as Qalandars train and exhibit "dancing" sloth bears as a way of life. The bears are captured as cubs, their teeth and claws extracted, and a rope or ring put through their nose. More than one thousand dancing bears are trained to entertain urban and tourist audiences in India, according to David Macdonald (2001). If dancing was all that was required, the sloth bear would be uncomfortable, but not in peril. Along with the Asiatic black bear, this species is so extensively collected throughout its range for its gall bladder and the bile therefrom for use in traditional Chinese medicine that it is considered an endangered species and classified by the IUCN as "vulnerable, close to extinction."

The Asiatic black bear (*Ursus thibetanus*) is black with a light muzzle and a distinctive chevron on the chest, often in the shape of a crescent moon, which gives this species its other name, "moon bear." The chevron can range in color from creamy white through lemon yellow, and in width from pencil-line thin to a boomerang-thick crescent moon.

Asiatic black bear (*Ursus thibetanus*).

These bears, which are frequently represented in Japanese art, have particularly large ears. The moon bear inhabits a considerable portion of Southeast and eastern Asia from Afghanistan and Pakistan and northern India through Nepal, Sikkim, and Bhutan, and into southern China, Thailand, Laos, and Vietnam; isolated populations are found on the northern Japanese islands of Honshu and Shikoku, and in the Russian Far East. Their range in these regions now typically comprises highly isolated and noncontiguous areas of land, all subject to human encroachment. An unanswered—perhaps unanswerable—question is why an animal with a coat better suited for cold, mountainous terrain should also inhabit dense jungles in Southeast Asia. Their large ears might be useful for dissipating heat, though that doesn't seem a strong explanation. The moon bear, in any case, has the unfortunate distinction of being the species most favored by the Asian medicinal market for its organs' potency. It has been devastated by poachers and is at risk of extinction in the near future throughout most of its range.

While it has been bears in Asia that have traditionally been hunted for their gall bladders, now, as their numbers have been reduced, bear species from all over the world—including polar bears, American black bears, and grizzly bears—are being killed to feed the needs of traditional Chinese medicine for bear parts. "Forty thousand black bears are killed legally in North America each year. An unknown portion of these animals parts are illegally traded on international markets," wrote Keith and Suzie Chang Highley in 1994. "Whole bear carcasses are now being found in the forests of Canada and the Soviet Union, with nothing but their gallbladders cut out," Judy Mills and Christopher Servheen tell us in their 1991 report on bear farming. In a 1994 Earthtrust report, "Bear Farming and Trade in China and Taiwan," Highley and Highley documented several instances of hunters and "businessmen" arrested for bear hunting in California and possession of bear parts, such as paws and gall bladders. In one such case, a California businessman was caught buying 164 bear gall bladders from undercover fish and game agents, and boasted that he had purchased on occasion three times that number. By 1998, in California alone, the illegal trade in bear parts was estimated to be worth millions of dollars a year (Phillips and Wilson 2002), and the problem extends way beyond California. As Judy Mills, Simba Chan, and Akahiro Ishihara wrote in their 1995 TRAFFIC report *The Bear Facts*:

> Today's market for bear gall bladders seems to be growing more prevalent outside Asia wherever there are bears. This trend is illustrated by reports from Ecuador, where Korean businessmen are said to be contracting *campesinos* to kill the relatively rare spectacled bear for its gall bladder. Sold at US$150, each gall bladder is worth five times the minimum monthly wage in Ecuador. Before development of this Asian market in South America, the spectacled bear already faced tremendous pressure from shrinking habitat and nuisance animal control.

In 1999, Sy Montgomery traveled to Southeast Asia to search for the golden moon bear, which she thought might be a species distinct from the ordinary black version of *Ursus thibetanus*, and later wrote of her investigation in *Search for the Golden Moon Bear: Science and Adventure in Pursuit of a New Species*. With zoologist Gary Galbreath (who had seen one in a cage in Yunnan, China, eleven years earlier), and a colorful cadre of local naturalists, drivers, and guides, she traveled through the zoos and jungles of Thailand, Cambodia, Laos, and Vietnam, seeking the golden bear. The group found many black bears in various captive situation, and on several occasions found themselves in the company of caged golden or light-colored bears. (In all the time they spent in the jungle, they never saw a wild bear.) They found bears in cages, parts of bears in markets, and stories confusing enough to raise their hopes that there might actually be a species of moon bear that had not yet been described by science. The book contains a collection of color photographs of light-colored and even *lion*-colored moon bears, and on the dust jacket, the author is shown with a living golden bear. Ever hopeful, Montgomery never actually relegates the idea of discovering the golden moon bear to the category of myth, but Galbreath, whom she often refers to as "the scientist," does not go beyond acknowledging the existence of a color difference.

Although lion-colored moon bears are indeed spectacular and worthy of a trip to Southeast Asia to investigate their evolution, other bear species are known to occur in assorted colors. The best known of these

When Sy Montgomery went looking for "golden moon bears" in Thailand, she found this spectacular lion-colored creature; alas, it was just a color phase, not a distinct species.

is the familiar American black bear, which comes in cinnamon, dark brown, grayish blue, silvery gray, reddish yellow, blond, and even white; there are sometimes different-colored cubs in the same litter. Brown bears, once separated into various species and subspecies, including grizzlies, Kodiak bears, and Eurasian bears, have now been lumped into one species—*Ursus arctos*—regardless of habitat or appearance. (Grizzlies get their name from their "grizzled" coat, which is often several shades of brown and silver.) The spectacled bear, sun bear, and sloth bear are usually black, but other color phases are occasionally seen. Polar bears, whose lives depend on being white, do not occur in other color phases, and giant pandas are always black and white. It's probably just as well that there isn't a separate species of golden moon bear; imagine how valuable *its* gall bladder would be in a Hong Kong pharmacy.

The use of bear parts in Chinese medicine dates back thousands of years. In the section on animal drugs from Li Shih-chen's 1597 materia medica *Pen Ts'ao Kang Mu*, we read that the Himalayan black bear (the moon bear) is

> as large as a boar, eyes vertical, humanlike feet, black, very fat in spring and summer, thick-skinned with tendons that stand out. It likes dropping down from trees. Traveling several thousand li it sleeps in caves and hollow trees which are commonly known as "bear's hotels." The gall is said to move to the head, belly and paws according to the season; similar to the elephant.
>
> BEAR'S GREASE. Sweet, slightly cooling, nonpoisonous. Used in lamps the smoke is weakening to the eyes, making them lose their lustre. Prolonged use as a food strengthens the mind, prevents hunger, lightens the body, gives longevity. To remove numbness and total loss of sensation. For feverish colds. Applied to blacken the hair, and promote its growth. . . . To cure baldness and ringworm. To remove pimples and blackheads from the face. A tonic and a cure for "wind" diseases, rheumatism.
>
> BEAR MEAT. Sweet, bland, nonpoisonous. Should not be taken if patient has a chronic disease. For rheumatism and weakness. The uses are the same as for bear's grease. For beri-beri with paralysis.

Bear's Paw. When eaten it keeps off colds and benefits the vitality.

Bear's Gall. The airdried material is used. It is so commonly adulterated that one should test it in water. One drop of the genuine article will give a line in the water which does not spread. [The line is often called the "thread of gold."] Drawn across a pool of ink, the ink should retreat form the tract. Bitter, cooling, nonpoisonous. For epidemic fevers, jaundice, chronic summer dysentery. For angina pectoris, ear and nose ulcers, and all evil sores. Anthelmintic [antiparasitic, e.g., worms]. Infantile convulsions. Antipyretic [fever-reducing]. It clears the mind, quietens the liver, and clears the sight. To remove pterygium [tissue growing over the eye]. For conjunctivitis, blindness in the newborn and various eye troubles. For caries.

Spinal Cord of the Bear. For deafness and giddiness. Rubbed on the scalp to promote the growth of hair and remove dandruff.

Blood of the Bear. For nervousness in children.

Bones of the Bear. For rheumatism of the joints and nervousness in children.

Offering cures for almost everything, the bears didn't stand a chance. When Zhang Enquin published the English-Chinese *Rare Chinese Materia Medica* in 1989, he wrote:

> If adulterated with yak biles there is a smell of seafood; if with sheep biles, there is a smell of mutton. . . . The bear bile (or kernel) tastes bitter first, sweet thereafter. There is a prolonged, cool, refreshing and tingling sensation on the tongue. It dissolves rapidly in the mouth, and gives a cool, refreshing sensation down the throat. It does not stick to the teeth when chewed. The counterfeits and adulterations have no fragrant smell, cool, refreshing sweet and tingling sensations to the tongue, but only bitter and fish-stench taste.

Once you are sure you have the genuine item with no fish-stench taste, you can use bear gall to "remove heat from the liver to relieve convulsion and spasm, treat infantile convulsion, epilepsy, hyperspassmia and eclampsia gravidarum [hypertension in pregnancy] caused by strong liver wind and extreme heat. It also improves vision and improves nebula, treats conjunctival congestion, swelling and pain in the eyes, photophobia and nebula caused by flaming up of liver-fire. It is efficacious in treating sores, furuncle [hair follicle infection], carbuncles, hemorrhoids, and sore throat." Some practitioners do not believe that the most effective medications have to come from endangered species. As Bensky and Gamble (1993) comment, "because of the high price of *Vesica Fellea Ursi* [bear gall bladder], often the gallbladder of the cow, *Vesica Fellea Bovus*, is substituted at a higher dosage. This practice is recommended because of the endangered status of many bear species."

Bear gall, like rhino horn, is not used as a love potion, although many in the West think it is. It may, in fact, be self-prescribed by some users for this purpose, but traditional medicine physicians consider bear gall to be one of the most powerful of all general medicines. It is most often prescribed for chronic illnesses of the liver, gall bladder, spleen, and stomach, but normally only after other gentler, less expensive herbal remedies in the "cold" category have failed to cool the" heat" of the disease (Mills and Servheen 1991). Its purposes are legion. In Wiseman and Ellis's 1996 *Fundamentals of Traditional Chinese Medicine* (the textbook used in some Chinese medical schools today), for example, bear's gall (*Ursi Fel*) is included in the category of "Heat-clearing, toxin-resolving medicinals" and is described as

> cold; bitter; nontoxic. Enters the liver, spleen, stomach and gall bladder channels. Clears heat, settles tetany, brightens the eyes and kills worms. Treats heat jaundice; summerheat diarrhea; child fright epilepsy; organ disease, hookworm infestation; eye screens; throat impediment; clove and malign sores; hemorrhoids. *Directions*: Oral: use in pills and powders. Topical: grind and apply mixed; use as eye medication.

Until 1984, bile for use in TCM was only obtained from wild bears killed for the purpose, but with bear populations tumbling, North Koreans developed the concept of extracting bile from living animals. The practice rapidly spread to China, and by mid-1992 there were reportedly 200 operating bile farms in China with about 4,000 bears. Later the farms were consolidated—often under government control—and by

1998 there were 247 farms holding 7,002 bears (Phillips and Wilson 2002). By 2003, the Vietnamese government acknowledged 2,000 wild-caught bears on farms in that country, but the total is probably higher.

The farm operation is typically set up in this way: Bears are caught in the wild and kept in squeeze cages so small they cannot stand and can barely move. Steel catheters are inserted into their gall bladders to drain the bile continuously into a plastic sac or a bowl, to be collected at regular intervals. The liquid bile is then oven dried to form crystals, which are used to manufacture various commercial products that range from powders, capsules, tonics, salves, and eyedrops, to teas, wines, and shampoos. When bears can no longer produce sufficient bile, they are usually put into another cage where they wait to die, or are killed for their paws and the gall bladder itself.

According to Dr. Gail Cochrane, a veterinarian with the Animals Asia Foundation, there are three methods of extracting bear bile:

1. The original technique employed latex catheters in which one end of a latex tube was surgically placed in the gall bladder. A

A moon bear in a squeeze cage at a Chinese "bear farm." The bear's gall bladder has been surgically exposed for catheterization.

metal tube with a plastic disk at one end was tied on and the structure held in the gall bladder with a purse-string suture. The other end of the latex was then threaded out through the abdominal muscle incision and passed up and over the flank under the skin to exit over the hip area. On exiting the skin, the latex tube was tied in a knot. To extract the bile, the knot was untied and a syringe was inserted into the tube and negative pressure applied.

2. A variation on this technique has the catheter threaded through the abdominal muscle and exiting in the abdominal area. The end of the catheter is then attached to an empty intravenous fluid bag, into which the bile will constantly drip. The bear is fitted with a metal corset and the fluid bag sits in a small metal box within the corset. To extract bile the bear must be anaesthetized, the metal box flapped open and the fluid bag changed. The "bile bags" are changed every one to two weeks.
3. Stainless steel catheters have replaced the latex catheters in most cases. A steel tube between 10 and 20 centimeters long, with a metal disk on one end (either cup shaped or flat), is equipped with a plastic tube. The disk end is surgically inserted into the gall bladder and secured with a purse string suture. The second disk generally lies just within the abdominal cavity, against the abdominal muscle. The remaining part of the catheter projects out of the bear's abdomen. Metal projections on the end of the catheter prevent the catheter from slipping inside the abdomen and deter the bear from chewing on the end. Frequently a piece of cotton wool or lint is inserted into the end of the catheter to prevent the bile from leaking out between extractions. Prior to extraction, the farmer removes this "bung" and either collects the bile by placing a dish underneath the catheter or, if the bile is not draining or draining slowly, the farmer will insert another thin tube through the catheter into the gall bladder to facilitate drainage.

The "Free-Dripping Technique" is the only extraction method allowed under the recent rulings from the Chinese Ministry of Forestry, which has jurisdiction over bear farms. A tube (technically a *fistula*) is created between the gall bladder and the abdominal wall, which involves either opening the gall bladder and stitching it directly to a corresponding hole in the abdominal wall or opening the gall bladder and creating a tube between this organ and the abdominal wall. The farmer inserts a catheter (either a rubber feeding tube or a stainless steel hollow probe) into the fistula once or twice per day to extract bile. If he does not do this regularly the fistula may heal over.

Among the first Westerners to visit Chinese bear farms were TRAFFIC investigator Judy Mills and her husband, bear biologist Chris Servheen. In 1990, they went to a farm in Sichuan, where they saw many bears in squeeze cages, but were not allowed to witness the "milking" process. In a 1991 article in *International Wildlife*, Mills observed that "some of the caged bears were so large they could sit up only in a slouched position. Other bears rocked their heads back and forth, or repeatedly threw their bodies against the bars of their cages. A two-year-old jumped up and down, banging its head." The investigators also went to a bear farm in Harbin, northern China, where once again, they were denied access to the extraction process. As they were leaving China and heading for South Korea, Mills observed:

> We saw bears in trouble; our hosts saw only dollars. We wanted conservation, they wanted medicine and profits as well as conservation, but the last only as a fringe benefit . . . black marketers in China offered to sell us bear gallbladders by the kilo or even wild bears whole . . . Clearly, bear farming does not prevent poaching of bears or trafficking in their parts. In fact, we found, farming bears for bile promotes the use of bears as a commodity and makes bear bile available to more people at lower prices.

Appalled at what they had seen, Mills and Servheen left China and headed for South Korea, because Mills said she was "determined to witness bear milking so [she] could tell the world what it looked like." In their 1991 TRAFFIC report, Mills and Servheen described the milking process on a farm outside Taegu in South Korea:

> Using a metal pole [the owner] prodded the bear into a narrow portion of its cage. As his wife distracted the bear with a pan of sweets, the door of the squeeze cage was lowered and metal rods inserted to

> confine the bear and keep its legs from interfering with its abdominal area. The owner reached in, unlocked the metal panel, and a plastic bag attached to a catheter dropped down. The bag was half full of a greenish brown liquid. The bear scraped and clawed wildly at the cage while the owner then extracted the liquid from the bag with an oversized hypodermic needle, withdrawing two full syringes. The process took approximately five minutes, after which the bear's tap was again locked behind the metal abdominal plate. The syringes were emptied into three plastic bottles, which were immediately packed in ice. A small amount was left so [the owner] squirted the contents into the client's mouths. The buyers paid about $1,700 per bottle, each of which probably held no more than 10 or 20 milliliters.

However awful it is for the bear, it is a boon to bear farmers; one couple running a farm in southern Guangdong said they made more than 300,000 yuan (US$51,600) in 1992. Advocates of bear farming argue that the practice helps to reduce pressure on wild bear populations, claiming that the gall produced by a single caged bear in a year is equivalent to the amount that could be harvested from forty wild bears which would have to be killed. Thus, the farmers argue, the benefits of bear farming are threefold: it protects bears in the wild, it produces revenue, and it provides valuable medicine for the treatment of human ailments.

Although the farmers' future looked bright, the same could not be said for the future of the bears. According to the World Society for the Protection of Animals, Chinese medicinal use of bear bile in the 1980s was 500 kilograms per year, but today consumption has escalated to around 7,000 kilograms per year for consumer products in China and abroad. It is estimated that there are no more than twenty thousand moon bears in China (Ma and Li 1999), and given the profits to be made by the capture and farming of bears, this species is profoundly endangered. Medicine shops in Japan, Indonesia, Malaysia, the Philippines, Korea, Hong Kong, Taiwan, Singapore, and Australia sell bear bile as well as other Chinese pharmaceuticals, as do shops in Montreal, Toronto, Vancouver, San Francisco, Chicago, Washington, D.C., and New York. In South Korea, bear gall bladders were fetching almost twenty times the price of gold (which was $11.53 per gram, or $326.29 an ounce) according to Mills and Servheen at the time of their 1991 report.

While traditional Chinese medicine is widespread in China, it is also popular in other Asian countries and elsewhere around the world. All eight species of bears are listed as endangered under CITES (Convention on International Trade in Endangered Species), and the Asian species are included in Appendix 1, which specifies that international trade in specimens of these species is, with few exceptions, not permitted. Because Taiwan is not recognized as a separate nation, trade between Taiwan and China is seen as a domestic issue and therefore does not technically fall within the purview of CITES. (Bhutan, Myanmar, Cambodia, North Korea, and Vietnam are also not CITES signatories.) Elsewhere, the export of bear bile products is illegal, but the ease with which the products can be smuggled or disguised has made it virtually impossible to curtail the trade, and dealers in "traditional" medicines around the world continue to sell bear products. For example, 681 kilograms of gall bladders, reportedly from sloth bears (*Melursus ursinus*), were exported from India to Japan between 1978 and 1988; because the average gall bladder can range in size from 50 grams to 125 grams (1.75 ounces to 4.37 ounces), this could represent the death of as many as ten thousand bears.

In many countries, however, bear parts are not exported at all, but used to satisfy a domestic demand. In his 1999 report on the status of black bears in India, Sathyakumar wrote:

> Black bear populations in India are largely threatened due to poaching for gall bladder and skin. While the former is believed to be of medicinal value, the latter is for trophy or ornamental purposes. The medicinal value of gall bladder is yet to be scientifically established, but tribes and local villagers strongly believe in its medicinal properties.

Wherever Asiatic black bears are found, people manage to find a rationale for killing them. In Japan, a country with no wildlife protection laws at all, wild bears are regarded as threats to crops, apiaries, fish farms, and livestock and may be shot, snared, or trapped with impunity. In Vietnam, on the other hand, the rationale is straightforwardly medicinal: "the bear's bile is the most appreciated because it cures many diseases, effectively treats the accumulation of blood flow below the skin, and counters toxic effects. Bear bone glue is used as a tonic, and bear fat is also a medicine and tonic" (Sam 1999). Mills and Servheen also documented the sale of bear gall and gall bladders in Malaysia, India, Nepal, Bangladesh, Indonesia, and Sri Lanka, which are CITES signatories, and Bhutan, Myanmar, Cambodia, North Korea, and Vietnam, which are not.

In 2001, the Animal Concerns Research and Education Society commissioned researchers to investigate the trade in bear gall bladder and bile products in Singapore. The researchers visited sixty-eight TCM shops out of a known eight hundred such establishments in the nation and found that fifty were selling bear bile products: pills, intact gall bladders, crystals, powder, and ointment. The price of an entire gall bladder ranged from 15 to 800 Singapore dollars (US$8.50–$440). It appears that Singapore, a nation famed for its strict laws and draconian punishments, is prepared to ignore malfeasance where traditional medicine is concerned. Vietnam has also been found to be a major offender.

Along with neighboring Cambodia and Laos, Vietnam once provided major habitat areas for both the Malayan sun bear and the Asiatic black bear. Illegal poaching has caused the numbers of both species to decline dramatically. The Vietnamese government responded by listing both species of bears as "rare and precious animals," protected by law from being poached, exploited, and utilized. Despite this legislative protection, both bear species are now facing total decimation as a result of the establishment of hundreds of illegal facilities throughout the country that keep bears for the extraction of bile. In Hanoi alone, at least seven hundred bears are kept in tiny cages in backyards of homes. Most have lost paws through being caught in snares crudely made from motorcycle brake cables. Trussed up with wire, these bears pour into the cities smuggled in the backs of vans to begin their lives of misery. Sources estimate that at least two thousand bears are being kept for bile extraction throughout the country—and the industry is still expanding.

Adult bears in Vietnam are dying slowly from bile extraction, while cubs too small to produce adequate quantities of bile are slaughtered for their whole gall bladders and paws. Animals Asia investigators have revealed two methods of bile extraction in Vietnam. In the first, bears undergo major abdominal surgery to remove bile from their gall bladders every three months. The surgery is crude and unhygienic and, according to the Vietnam government, the bears usually suffer four such operations before dying from the infection and pain. Another method entails the extraction of bile with the assistance of an ultrasound machine, catheter, and medicinal pump. The bears are drugged, restrained with ropes, and their abdomens are repeatedly jabbed with 4-inch needles until the gall bladder is found. One operator was even witnessed licking the needle between numerous insertions in an attempt to locate the bile. The process often leads to dangerous leakage of bile into the body and a slow and agonizing death from peritonitis. Despite the laws, newspapers in Vietnam openly advertise bear bile, while restaurants and cafes throughout the country freely serve bear bile wine, bear paws, and bear meat. Animals Asia investigators have witnessed the carcasses of whole bears floating in glass tanks filled with fermenting wine and have filmed cages in restaurants containing distressed and dying adults and cubs awaiting their fate.

While bear bile has been used in traditional medicine for millennia, many Asian doctors agree that it can be replaced with herbal and synthetic alternatives. However, the Vietnamese community has fallen victim to a public misinformation and marketing campaign that promotes expensive bear bile as "the people's medicine"—a cure for cancer, impotency, and even hangovers. Under the direction of Jill Robinson, Animals Asia, which provided much of the basis for this account of the bear bile situation in Vietnam, has been working in that country since 1998, encouraging the Vietnamese government to enact comprehensive legislation to protect both wild-caught and captive-bred bears from the practice of bile farming. In 1999 the organization submitted a written proposal to the Vietnamese government offering to assist in finding solutions for bears currently held in illegal facilities. After two years of lobbying embassies and nongovernmental organizations in Vietnam for support, Animals Asia submitted official letters to Prime Minister Phan Van Khai and other government officials, encouraging them to act to end the illegal bear bile industry. In September 2002, a historic ruling by the prime minister and the government of Vietnam granted Asiatic black bears the country's highest level of protection, thus making the hunting, keeping, and exploitation of all bear species for the bile industry illegal under all circumstances. This marks a victory for Animals Asia—as long as the bear farmers respect the new legislation.

The Bear Bile Business, published by the World Society for the Protection of Animals (Phillips and Wilson 2002), is a 248-page document that contains detailed information on every aspect of the "business," from its history and profiles of Chinese bear farms to assessments of individual markets in various countries, including Canada, the United States, Indonesia, Japan, Malaysia, Singapore, Taiwan, and Australia. The section on Japan, written by Kumi Togawa, Masayuki Sakamoto, and assisted by Chie Iijima, begins:

> Japan is one of the biggest consumer countries of bear gall and bile—there is virtually no regulation in the current Japanese legislation to restrict domestic trade in bear gall bladders. There are 2,000–3,000 brown bears in Hokkaido, the northern Japanese island, and about 7,000 black bears in the other areas of Japan. Populations of both species are declining as their habitats are destroyed and fragmented. In Japan, over a thousand bears are killed annually for sport hunting and pest control, without the implementation of proper conservation control measures. In bear parks, bears are kept in inappropriate conditions, and some parks sell bear products, including gall bladders.

From 1979 to 1988, TRAFFIC Japan estimated that China exported between 11,000 and 59,000 gall bladders to Japan, and another 1,051 kilograms of gall bladders were exported from China to Japan between 1988 and 1990 (Phillips and Wilson 2002). Togawa, Sakamoto, and Iijima reported that "the demand for bear bile in Japan is still continuing at a level of at least 200 kg [440 pounds] per year. If we assume 20 g of dried bear gall is obtained from one bear, then in theory 10,000 bears must be killed to satisfy the demand. Even the farmed bears in China don't live very long, and the captive stock needs to be supplemented with wild caught bears."

"Japanese medicine"—which is not the same as medicine in Japan—is known as *kanpo*, which literally means "Chinese method." It is a somewhat altered version of TCM, and its pharmacopoeia includes bear gall, known locally as *yu-tan*. According to Mills and Servheen's 1991 overview, "bear gall is used in Japan for abdominal pain, fever, liver detoxification, dyspepsia, nausea, poor appetite, skin burns, weakened heart, infant colic, every kind of intestinal disease and more." The collecting of bear gall and bear parts—including gall bladders—is legal in Japan, and Japanese pharmaceutical companies are able to sell bear gall as a legitimate medicine.

Despite the exploitation of the animals for bile, bears remain popular subjects in Japanese art, and some bears, particularly moon bears, are still to be found wild on the islands of Honshu and Shikoku. Live specimens can be seen in zoos and in "bear parks," popular attractions that feature only bears. As of 2000, there were eight bear parks in Japan and two more under construction. Some feature all species of bears except the giant panda. The bears ride bicycles, jump through flaming hoops, roller-skate, play basketball, and dance, and they are often dressed in funny costumes. When the bears outlive their usefulness as performers or when they die of natural causes, the parts are sold, with a gall bladder bringing as much as $4,000 from pharmaceutical houses. "In early 1991," wrote Mills and Servheen, "more than 100 bears from one Japanese bear park were selected for slaughter. After the bears were rendered at a wildlife-meat packing plant, all of their gall bladders were sold to a South Korean broker. To date, more than 1,000 bears have died or been slaughtered in this park."

Once upon a time, the forests of Korea were home to Asiatic black bears, called *Pandalgom* by Koreans, but hunting, deforestation, and wars have taken their toll. Korea's wild bear populations, at best, have been reduced to single-digit figures. Some Pandalgom are exhibited in zoos, but when they become too old to entertain visitors, they are sold at public auction for their much-sought-after gall bladders. According to Mills and Servheen (1991), in 1990, the going rate for a bear was about $7,100. "Koreans are perhaps the most dedicated of all Asians to the use of bear gall bladder as medicine—more so than the Chinese, who originated the practice. Some Koreans are willing to pay more for bear gall and go to greater lengths to get it than people of any other nationality." Like traditional Japanese medicine, Korean medicine is based on centuries-old Chinese precepts and is said to have arrived in Korea at the time of the Han Dynasty, which dates from 206 BC to AD 221. By official decree, Western medicine was kept out of Korea until 1884, but under Japanese colonial rule (1910–45), it was reintroduced and soon became the dominant practice. After the Korean War (1950–53), there was a resurgence of traditional Korean medicine and even as South Korea was becoming highly industrialized, there was a flowering of interest in ancient traditions such as ancestor worship, filial piety, tonic foods, and herbal medicine (Mills and Servheen 1991). Of traditional Korean medicines, bear gall (*ungdam*) is considered the most powerful, able to purge all toxins, cure liver ailments, and treat diabetes, high blood pressure, palsy, fever, and hemorrhoids.

In South Korea, street peddlers of Chinese medicines were ubiquitous; Mills and Servheen found that "medicines such as bear gall can sell in the streets of Seoul for 10 times their price in China. . . . Among the ethnic Koreans' offerings we counted 136 bear gallbladders, several Asiatic black bear paws, vials of bear bile crystals, and numerous medicines containing bear gall. One vendor offered to trade a bear gall for our Nikon camera . . . we were offered two bear gallbladders for $2,861 by one vendor. Another wanted $700 for a single gall and $60 for a one- to two-gram vial of bear bile crystals. Yet another vendor showed us a very large bear gall for which he wanted $3,292." Where did the bear gall come from? Korea itself was no small supplier: As of 1990, there were fourteen bear farms in South Korea holding a total of 655 bears.

Relation to CITES regulations, as noted above, is different for the island of Taiwan than it is for other areas of Asia. As recently as 1949, the island of Taiwan (then known as Formosa) was considered part of China, but after Mao Zedong took control of the mainland, the Nationalist Party led by Chang Kai-shek broke away and set up its own Republic of China on Taiwan. There can only be one "China" in the United Nations, and most countries recognize the huge People's Republic on the mainland. Taiwan, therefore, is not eligible to join the United Nations, and it is a pariah nation as far as international treaties such as CITES are concerned. While many nations circumvent CITES restrictions on importing or trading in endangered species, Taiwan ignores them completely. Although there are government restrictions in Taiwan on hunting moon bears, these too are ignored, and gall bladders, bile, and bear paws are easily found in "game shops" and pharmacies. In the capital city of Taipei, TRAFFIC investigators Mills and Servheen found gall bladders selling for anywhere from $363 to $1,454, bear meat for $34 per kilogram, and paws for $181 to $363. The average price for a whole bear was $2,713.

Taiwanese medicine is little different from traditional Chinese medicine, although Western practices are said to be less integrated than they are on the mainland. This places an even greater emphasis on tonics and herbal cures, and topping the list are bile and other bear gall products. In their report on bear markets in Taiwan, Chen, Wu, Bhiksu, and Fisher estimated that there were ten thousand TCM shops on the island. Taipei's Di Hua Street is notorious for its profusion of Chinese apothecaries, and of the thirty-four medicine shops they investigated, Mills and Servheen found that thirty sold bear gall bladders, one with as many as twenty-five in a single tray. Those shopkeepers willing to talk said that most of their material came from the bear farms of mainland China. If that is true, political differences between China and Taiwan clearly do not interfere with the lucrative bear bile business. Often mixed with ground pearls, cow gallstones, musk, and amber, bear gall is a popular and powerful tonic for the heart, lungs, stomach, and kidneys among Taiwan's population of 22 million and is considered good for the skin as well. All shopkeepers said that bear bile was used in a rite for newborn babies, where a fraction of a gram is placed on the baby's tongue to cleanse the blood of poison passed by the mother to the child in the womb. "Sometimes," Mills and Servheen noted, "rhino horn, gazelle horn, saiga horn, coral, dried palm ginseng, dried insects, stalactites, oxidized mercury, and even gold are added for a more expensive and powerful tonic." The United States invoked the Pelly Amendment to impose trade sanctions on Taiwan in April 1994, and went on to announce a ban on imports of wildlife and wildlife products from Taiwan, effective August 19, 1994. Two years later, on September 11, 1996, recognizing Taiwan's attempts to control the illicit trade in bear products, the United States announced that Taiwan was being removed from the Pelly Amendment's "Watch List."

Beginning with the promulgation of the *Wildlife Conservation Law* in 1989 and continuing through 1995 with the formation of a task force designed to investigate and supervise wildlife conservation and crack down on the smuggling of wildlife products, Taiwan began to demonstrate a commitment to domestic conservation and support for global wildlife protection efforts. In the 1999 IUCN report *Bears: Status Survey and Conservation Action Plan* (Servheen, Herrero, and Peyton, eds.), there were indications that Taiwan was working to overturn its reputation as a primary consumer of illegal bear parts, but as Ying Wang, a bear biologist at Taiwan Normal University, suggested in that volume, "Though the sale of bear parts and meat has been ended officially, it still exists on the black market."

TCM has long depended on drinks, pills, ointments, and powders that can be varied according to the diagnosis of the individual. New to Taiwan is scientific Chinese medicine (SCM), which resembles Western medicine in that the ratio of ingredients in a particular potion or ointment is fixed, and one product—such as aspirin or ibuprofen—fits all diagnoses. SCM is less expensive than TCM because it does not require that a potion be mixed but rather can be bought right off the shelf. Unregulated SCM can, of course, use genuine bear bile, but it is easier for unscrupulous dealers to substitute cow or pig bile—or use no bile at all.

Laos is not a signatory to CITES and, like Taiwan, overtly flaunts endangered species strictures. The former director of the Forest Resources Conservation Project served time in prison for trafficking in Laotian and Thai wildlife, according to Mills and Servheen's 1991 report. After the fall of Saigon in 1975, Laos fell into the hands of Communists, who proclaimed it the Lao People's Democratic Republic. That the heavy-coated *Ursus thibetanus* can still be found in Laotian jungles is more a testimony to the density of the underbrush than to any desire on the part of Laotian authorities to protect the bears. Her quest for golden moon bears took Sy Montgomery to Laos, and she found several bears (and a young tiger) that had been caged in private homes and in the zoo at Vientiane.

"It seemed no species in Laos was exempt from human hunger for food or for money," Montgomery notes. The Laotian people use bear gall bladder as medicine, Mills and Servheen tell us, and "they also use bear fat to treat rheumatism and eat bear meat for energy. Medicinal use of bear parts is illegal, but enforcement in this regard is mostly nonexistent." Although undocumentable, it has been said that a large proportion of bear parts that end up in Korea originate in Laos. Because the Mekong River forms the border between Laos on one bank and Thailand and China on the other, smuggling and illegal trade are rampant. Today, Laos is one of the ten poorest countries in the world, and valuable bear parts can take the place of worthless currency.

Montgomery also saw captive moon bears in Thailand, at the Banglamung Wildlife Breeding Center outside Pattaya, at the Lop Buri Zoo, and at the Million Year Stone Park outside Pattaya. Compared with other Thai bears, these were the lucky ones. "So prized are bear parts," Montgomery wrote, "that forty live sun and moon bears were smuggled out of Thailand into South Korea, where their meat, blood, and gallbladders were used to fortify the Korean 1988 Olympic team. So effective are bear remedies thought that Hyundai Corporation's elderly ex-president, Jung Ju Young, publicly credits his excellent health to regular supplements of bear bile."

Thailand became a CITES signatory in 1983, but to date, there is no national legislation in place to enforce the regulations. In Bangkok's Chinatown, Mills and Servheen readily found bear gall bladders for sale and learned that there were restaurants that specialized in serving bear meat and bear paws. In 1991, the new government decided to put an end to the enormous trade in flora and fauna that had made Bangkok the epicenter of smuggling activities in Southeast Asia. Bear farms and restaurants were shut down and their proprietors arrested, but with the high prices bear parts in Korea, China, and Japan command, it is highly unlikely that the bears of Thailand will be granted a reprieve.

But it is not only Asian countries that import and use bear bile medicines; any country with a substantial population of Asians (often congregating in "Chinatowns") will manifest an interest in TCM. For the *Bear Bile Business*, the World Society for the Protection of Animals sent investigators to three cities in Canada (Vancouver, Toronto, and Montreal) and to four cities in the United States (San Francisco, Washington, D.C., Chicago, and New York), where a total of sixty-five TCM shops were examined. Fully 78 percent (fifty-one out of sixty-five) sold bear gall bladder or bile products. Bile medicines, usually containing bear bile combined with other herbal medications, were found in the form of oils, ointments, pills, and plasters and are usually used for rheumatism, hemorrhoids, and sprains. Bile powder, made from bile collected on Chinese bear farms, is sold as flakes and packaged in vials. Seventeen intact gall bladders were found in five shops, collected from Chinese, Russian, and North American bears. One gall bladder in a Toronto shop was priced at $650.

Researchers in Australia, where there are a substantial number of people of Asian descent, had similar experiences in the Chinatowns of Sydney, Melbourne, and Brisbane. Pei-Feng Su, Janey Wong, and Yi Chiao surveyed thirty-five shops and found that seventeen (49 percent) sold bear bile products (the most popular of which was Fel Ursi Hemorrhoid Ointment); one shop in Sydney had six gall bladders for sale. There are no bears in Australia (koalas are marsupials, not bears), so all bear products must be imported and are often smuggled in by individuals and then sold commercially.

The active ingredient in bear bile is ursodeoxycholic acid (UDCA). A Google search for "Ursodeoxycholic acid" produced forty thousand hits and revealed that it is actually used in Western medicine in an attempt to cure some medical problems, usually liver- or gall bladder-related: UDCA is a naturally occurring bile acid found in small quantities in normal human bile and in large quantities in the bile of certain species of bears (not including the giant panda, which has enough troubles of its own). It is a bitter-tasting, white powder freely soluble in ethanol and in glacial acetic acid, slightly soluble in chloroform, sparingly soluble in ether, and practically insoluble in water. With trade names such as Actigall, Arsacol, Cholit-Ursan, Destolit, Deursil, Litanin, Ursodiol, Ursochol, Ursofalk, Ursolvan, and Ursotan, it is prescribed for the dissolution of gallstones, but the treatment requires months of therapy; complete dissolution does not occur in all patients and recurrence of stones within five years has been observed in up to 50 percent of patients. UDCA is also prescribed for treatment of cirrhosis of the liver, and, according to results of a follow-up study of a randomized trial published in the April 2003 issue of *Gastroenterology*, it protects against colon cancer in patients with ulcerative colitis and primary sclerosing cholangitis (where the bile ducts inside and outside the liver become inflamed and scarred, and the ducts become blocked). UDCA can be synthesized for about 16 cents per pill, but practitioners of TCM prefer to obtain it from bear gall bladders.

Does bear bile, as prescribed in TCM, actually do what it's said to do? Despite the fact that bear bile does, in fact, contain a significant amount of ursodeoxycholic acid, which, when synthesized from cow bile, is used to break down gallstones, John Seller of CITES points out that fakes in the market raise the question as to whether bear products themselves are efficacious. One study of samples of supposed bear bile turned out to be domestic pig bile. Dr. Lee Hagey of the Department of Medicine at the University of California observed that "pig is an effective substitute to bear bile as the crystals mimic bear crystals and the gall bladders look the same. I believe that probably many people in Asia have been taking pig bile and not bear bile. I also believe that before the current fad of bear bile, Asians used pig bile labeled bear bile. So we can say that they have always used pig bile with good effect."

By 1991, Judy Mills had published her exposé of Chinese bear farms in *International Wildlife*, in which she observed that "more bears live on bile farms than probably remain in all of China's forests. Yet the Chinese government has asked its wildlife specialists to concentrate their energies on boosting farmed bear birth rates, bile outputs, and profits." It was during the search for additional information on bear farming that I came across the work of Animals Asia. I was astonished to find an organization in China (Hong Kong, actually) doing so much to improve the terrible plight of farmed bears in China. Jill Robinson, the organization's founder, describes her introduction to bear farms in China:

> Sometimes we receive a message in life which is hard to ignore. For me, that message came in 1993 when I walked onto a bear farm in China for the very first time. Nothing prepared me for that moment and it was with utter disbelief that I witnessed a scene which would subsequently change my life and which would start the dream of the China Bear Rescue. . . . Bear farming was virtually unknown in the

> West and it was only when I heard rumors of a bear farm operation across the border in southern China, that I joined a tour group from Hong Kong to witness the practice at first hand. Whilst the bear farmer and his wife proudly demonstrated their bile preparations, I stole away from the group and found some stairs leading to a room below. As my eyes became accustomed to the darkness, it was as if a horror story was unfolding itself frame by frame. Row after row of tiny wire cages held living, breathing bears as prisoners—bears, I was later to discover, which had spent 13 years of their life behind bars. Resembling victims of medieval torture, these pitiful animals turned around to reveal infected, gaping wounds in their stomachs, from which protruded rusting, metal catheters. . . . At one point I felt a gentle tap on my shoulder and turned around to see a female moon bear reaching out through the cage. Without thinking, I took her paw and, whilst gazing into sad, dark, unblinking eyes, made a pledge that one day I'd be back to set her free. (Animals Asia 2003)

Born in the United Kingdom, Robinson arrived in Hong Kong in 1985 and spent the next twelve years working in Asia as a consultant for the International Fund for Animal Welfare (IFAW). In 1998 she founded the Animals Asia Foundation and focused on the practice of bear farming. To date, Animal Asia's China Bear Rescue has rescued 185 bears in China and is working to end the practice of bear farming by the 2008 Beijing Olympic Games. In 1995, Robinson was presented with the Readers Digest "Hero for Today" award and in June 1998 she was a recipient of the Member of the British Empire award, presented by Queen Elizabeth on the Birthday Honors List in recognition of her services to animal welfare in Asia.

In 1993, after her visit to two bear farms, one near Hong Kong, Robinson launched a wave of media articles condemning the farms and the treatment of an estimated ten thousand or more bears for the bile business. The organization was able to force the closure of the two farms and to rescue and provide medical care for nine bears from one of the farms. Animals Asia Foundation assumed responsibility for the bears' rescue and subsequently began a dialogue about bear treatment with Chinese authorities. David Chu Yu-lin, a prominent Hong Kong businessman and politician, and a hunter until his children persuaded him otherwise, played a crucial role in this campaign. Making use of his contacts as an advisor on Hong Kong affairs to China's State Council, Chu Yu-lin took the cause to high levels in the Chinese government, where the plight of the bears was officially acknowledged.

In November 1994, the China Wildlife Conservation Association, the IFAW, and two Hong Kong based groups—Earthcare and the Chinese Association of Medicine and Philosophy—signed an agreement to work together to end bear farming. The immediate goal was rapid closure of the most abusive farms and the long-term goal the total elimination of bear farming, providing alternative medicines could be found.

Most bear farms are located in Sichuan Province in south-central China. In 2000, an agreement between Animals Asia and China Wildlife Conservation Association and the Sichuan Forestry Department was reached that would close the worst farms, release five hundred bears into sanctuaries, and begin the process of eliminating bear farming

Signing the historic agreement to release five hundred moon bears into the care of Jill Robinson, who is shown with officials of the China Wildlife Conservation Society and the Sichuan Forestry Department in 2000.

in that region. Completely. The agreement was sanctioned by the Central Government of China, and Animals Asia agreed to pay each farmer a sum of money for each bear released into its care. It costs about $600 per year to feed each bear in the sanctuary and another $2,000 per bear per year to cover the costs of workers' salaries, medical and surgical supplies, and construction costs at the sanctuary. Donations to Animals Asia are not only necessary—they are critical to the care of bears.

In October 2000, Animals Asia received the first group of pitiful bears. Since that time, thirty-eight farms have closed and as of January 2005, Animals Asia has received 185 moon bears. The bears typically arrived bone thin, desperately ill, and terrified, the victims of catheter implantation or the newer "free-dripping" technique that creates a permanent hole in the gall bladder—an open wound responsible for the high mortality rate on the farms. Many of the bears are missing limbs from being caught in traps in the wild, while others have had teeth, claws, and paw tips hacked away to make the bears easier to handle. Animals Asia veterinarian Gail Cochran euthanized several bears as a result of peritonitis, septicemia, and massive abdominal or facial cancers. However, 120 bears are now enjoying their freedom in the "China Bear Rescue" forest sanctuary in Chengdu in central China.

Shortly before the bears were scheduled to move to the sanctuary, the IFAW gave US$75,000 to China's State Administration for Traditional Chinese Medicine (SATCM) to support research to develop a single herbal alternative to bear bile. Like many others in the traditional Chinese medicine community, SATCM staff were unhappy that Chinese medicine was blamed for the endangered status and cruel treatment of some wildlife species. They argued strongly for legitimate medical use of bear bile and other animal products while acknowledging that much of the international trade in animal parts resulted in unconscionable profits to unscrupulous middlemen. To He Huiyu, deputy director of SATCM's Department of Science, Technology, and Education, the grant award signaled that the end of bear farming was near. He anticipated that acceptable alternatives would be ready in about four years. In a 2002 article, Dr. Peter Li of the University of Houston wrote, "International efforts to end bear farming have unavoidably touched two most sacred objectives of contemporary

China: modernization and cultural revival. The Animals Asia's experience has shown that these two issues do not constitute insurmountable obstacles to international efforts to create positive change in mainland China."

In September 2003, an article about Jill Robinson's campaign appeared in the *National Enquirer*, a well-known American celebrity gossip tabloid. Sandwiched between stories about Ben Affleck and Jennifer Lopez's failed attempt at nuptials and George Clooney's shopping habits is a full-page story entitled "Woman Helps Bears Claw Their Way to Freedom." It begins, "Horrific cruelties inflicted on captive bears was more than Jill Robinson could bear—so she's started a miraculous sanctuary that's saving hundreds of these tortured animals." The article, which is unsigned, includes photographs of caged bears as well as pictures of the rescued bears at Chengdu. When I asked Annie Mather, media director for Animals Asia, about the choice of *National Enquirer* as a place to publicize the bear rescue, she said, "Yes, indeed, I am well aware that the *Enquirer* is the most notorious of tabloids (with a huge circulation)—however, because it is a nice article and lists our Web site, we don't believe the article is giving the

Members of the Animals Asia staff sitting atop some of the cages from which the bears were released.

wrong impression of our work—so far, we are getting supportive e-mails from people who have never heard of our rescue before."

Traditional medicine practitioners are beginning to recognize the problems inherent in the international trade in animal parts, and in response, some are adopting different approaches to treatment, including herbal alternatives. In a 1993 paper, for example, representatives from the Chinese Association of Medicine and Philosophy presented a list of herbal substitutes for rhino horn, bear gall, and antelope horn. For rhino horn they recommended herbs by the name of *Rehmannia glutinosa* and *Captis chinensis*, which demonstrated significant antipyretic activity during a test conducted by Professor Paul But of the Chinese University of Hong Kong. For antelope horn they recommended *Tian Ma* (gastrodia root), *Gou Teng* (uncaria), and *Ju hua* (chrysanthemum), all of which are identified by Wiseman and Ellis as "yang-subduing wind-extinguishing agents." In 1994, the Chinese Association of Medicine and Philosophy joined with Earthcare to produce a report entitled

Some of the bears lost limbs in the traps that were used to capture them. Here "Three-Legged Andrew" recuperates at the sanctuary established by Animals Asia after having been released from his cage.

"Herbal Alternatives to Bear Bile." For every ailment currently treated with bear bile, they argue, there is a combination of plant-derived, nonendangered alternatives that are cheaper, more readily available, and just as effective.

According to Animals Asia's 2002 newsletter, UDCA was first synthesized in 1954 from chickens and has proven to be efficacious in treating liver disease. Ironically, far more synthetic bile is consumed in Asia than bear bile—Japan, Korea, and China consume 100 tons of synthetic bile annually. Synthesized UDCA (using cow or pig bile—or even *no* animal products) is a medicine that seems to have been successfully used worldwide to treat gallstones, primary cirrhosis, autoimmune hepatitis, and colon cancer. Research in the United States in 2002 has shown that bile acid is able to cross the blood-brain barrier and may be beneficial in treating Parkinson's, Huntington's, and Alzheimer's diseases. Researchers stress that this bile acid can be produced synthetically, but for some, artificially produced substitutes may not be satisfactory.

Though suitable herbal and synthetic alternatives exist, the problem of bear exploitation is far from solved. Human greed, difficult to legislate against, is still powering the bear farming industry. CITES restrictions in some cases only force the industry underground and encourage the growth of a black market. Groups like the World Society for the Protection of Animals, however, have mounted campaigns in Hong Kong, Singapore, and Japan to bring the issues before the public and to show that there are viable alternatives to keeping catheterized bears in tiny cages. According to Duncan Graham-Rowe (2004), "Customs officers will soon be armed with bear detection kits in a bid to halt the burgeoning illegal trade in bear parts such as gall bladders and bear bile." Developed by the World Society for the Protection of Animals, the kit is based on the principle of home pregnancy kits; it uses antibodies to detect proteins that are specific to each bear species. Instead of the lengthy laboratory testing process, the kit will identify bear products immediately, and while it will obviously not eradicate the trade, it will be a deterrent to smugglers.

The publicized efforts of people like Jill Robinson to free caged bears bring to the attention of the world the inhumanity of bear farming. More than winning the release of five hundred bears from their torture

Portrait of a happy bear. "Jasper" at the Chengdu sanctuary.

chambers, Robinson's efforts serve to show that tigers, rhinos, sea horses, sea lions, and other species endangered by the needs of TCM don't need to be. Herbal or other remedies can be substituted; public education can alert potential consumers that the production of such medicines contributes to the extinction of rare and endangered species.

7

Tigers, Rhinos, and Bears—Oh My!

In 1993, the Chinese government published a "Notice Promulgated by the State Council on the Prohibition of Trade in Rhinoceros Horn and Tiger Bone," which banned the import and export of these substances in China. The notice included this wording:

> It is forbidden to sell, purchase, carry or mail rhinoceros horn or Tiger bone. The rhinoceros horn and Tiger bone, presently kept in stocks shall be examined, re-registered, sealed up and properly kept, and the owners shall declare their stocks accurately to forest departments at provincial levels . . . . The forestry departments at provincial level or the agencies designated by them shall prepare documentation of their stock of rhinoceros horn and Tiger bone and submit the documentation to the Office of the People's Republic of China for the Administration of the Import and Export of Endangered Species.

An enormous country with an enormous population, China alone can make a mockery of such "notices" if they're not adhered to. Besides, no notice by itself—no matter how repressive the government that issues it—can possibly squelch a people's belief in the efficacy of centuries-old miracle cures. It is clear that there are still many, many people, in China and elsewhere, who believe that tiger bone, rhino horn, and bear bile can cure everything from hemorrhoids and gallstones to cancer and impotence.

In 1997, TRAFFIC commissioned Judy Mills to learn what had resulted from the 1993 ban, and from 1994 to 1996, she sent Mandarin-speaking investigators to pose as customers at various medicine markets and pharmacies in China to see if there were any discernible trends in the tiger-bone and rhino-horn trade. Sales were down, but far from eliminated:

> Taken together, the results of these surveys could indicate that China has been highly successful in implementing the domestic ban on trade in rhinoceros horn, Tiger bone and their medicinal derivatives. Lack of pre-ban surveys of the trade prevent this or any other conclusions about availability before versus after the ban. What is more important, and of immediate conservation concern, is the fact that even a low level of availability exists in the world's most populous country—a country that depends, at least in part, on TCM to provide health care to 1.3 billion people. If poaching stands behind the source of rhinoceros horn and Tiger bone in China, the world's remaining rhinoceros and Tiger populations could not supply even a small residual demand in that country for long. (Mills 1997)

A TRAFFIC survey, conducted in 1996, found that about 7 percent of Hong Kong's adult population used TCM regularly and that users are more likely to be women than men. Most users of TCM were not particularly interested in the ingredients contained in the medications prescribed, but a significant proportion of those who knew they were using rhino horn or tiger bone said that they would continue to do so even if it was against the law. As one might expect, those who used TCM medications that contained wild animal parts were less inclined to support conservation measures, while those who didn't use animal-based TCM medications were more in favor of protecting wildlife. One-third of Hong Kong's adult population reported that they had consumed exotic animals (by our standards, anyway), with snake being the most popular exotic food item, and more than half of the population said they used tonics containing wild animal derivatives. A majority of Hong Kong Chinese, and especially those who did not use TCM, expressed concern about wildlife conservation and indicated they would voice support if they were informed of the relevant issues. Like most people, users of TCM medications in Hong Kong believe that their own health takes precedence over the health of an endangered species.

Some of the animals whose parts are used for TCM—particularly rhinos and tigers—qualify as endangered species. If a species is sufficiently endangered, there is a distinct possibility that it might become extinct. With regard to rhinos and tigers, we must therefore ask if stockpiling horn or bones might be financially advantageous—at least in the short term—if the animals themselves actually became extinct. The hoarders of horn and bone could then demand astronomical prices for what remains—as Will Rogers answered when asked how to make money: "Buy land, they ain't making any more of it." The idea that anybody would be pro-extinction is painful, but when Cory Meacham was writing *How the Tiger Lost His Stripes*, he interviewed Paul Goddard, director of the National Fish and Wildlife forensic laboratory in Ashland, Oregon, who indicated how plausible the idea must be. Speaking of confiscated tiger bones, Goddard said: "If I were doing this and had no scruples and I liked money a lot, I would, at whatever price necessary, buy up the bones of tigers and then pay poachers to kill the rest." Says Meacham, "This may sound like a far-fetched plot in a spy novel, but evidence of exactly such commodification is readily available for other endangered species. Massive stockpiles of rhino horn have been discovered, along with anecdotal reports from poachers claiming to have been instructed to kill rhinos in the wild whether they have valuable horns or not."

The slaughter of rhinos and tigers for the questionable needs of TCM is terrible enough, but it would require a view of humanity far more cynical than mine to see those who would encourage poachers as actually wanting to bring about the extinction of the species. I know of two instances, that of the quelili and that of the Tasmanian tiger, in which a species was made extinct by hunters when extinction itself was not the goal. Endemic to Guadalupe, an island 140 miles west of Baja California and 180 miles south of San Diego, was the quelili (*Polyborus lutosus*), a large bird of prey which was believed by the islanders to be taking young goats. The quelili was a species of caracara, a hawklike bird that looks more formidable than it is. The introduction of goats was probably beneficial to the carrion-eating quelilis, and in 1876 the birds were said to be abundant on every part of the island. By 1906, however, the birds had been so efficiently eradicated by the goatherds that not one remained alive. The Tasmanian tiger, or thylacine (*Thylacinus cynocephalus*), was a German shepherd–sized marsupial carnivore, striped on its hindquarters and tail, that was believed by Australians to be a sheep killer. Under a government-sponsored bounty system, they were relentlessly slaughtered; by the 1930s, they were all gone. Although both the quelili and the thylacine are extinct, it is unlikely that the men who killed them had this goal explicitly in mind; they just wanted to reduce the numbers of predators and protect their valuable livestock.

As with Jill Robinson's attempts to rescue bears in Asia and seek herbal alternatives for bear bile, there is a modicum of good news for the rhinos and tigers, as well as other species endangered by traditional Chinese medicine's demand for animal parts. The American College of Traditional Chinese Medicine and the World Wildlife Fund have formed a dynamic partnership to build public support for the conservation of tigers, rhinos, and other endangered species by reducing reliance on endangered species parts. In 2003, the World Wildlife Fund developed a Web site (www.tcmwildlife.org) that specifically emphasizes the plight of rhinos, tigers, and bears:

> Although tiger bone, rhino horn, and bear gallbladder use goes back at least a thousand years, illegal trade and poaching of these endangered and threatened species have increased significantly in the last two decades. The booming economies and growing wealth in parts of Asia have caused demand and prices to rise for many wildlife products. A major cause contributing to the ongoing depletion of tigers, rhinos, and bears is the use of their parts for traditional medicinal purposes.

In 2004, TRAFFIC North America published a study that demonstrated that increased awareness of animal endangerment could play a significant role in sales of TCM products. In a comparative study of traditional medicine markets in San Francisco and New York, investigator Leigh Henry found that New York far outstripped San Francisco in the availability of medicines containing tiger bone, leopard bone, rhino horn, musk, and bear bile:

> The differences in availability of medicines labeled as containing protected or regulated species in the San Francisco and New York City areas are alarming. Only 3% of San Francisco stores offered any medicines labeled as containing tiger, while 41% of New York City shops did. Rhino horn medicines were not found for sale in San Francisco, while they are available in 7% of New York City shops. Musk medicines were offered for sale in 58% of San Francisco stores and all New York City stores; and bear bile medicines were found in 24% of San Francisco shops and 70% of New York City shops.

The investigators found that San Francisco shopkeepers were "acutely aware of the laws regulating these species, and many also demonstrated that they were aware of the reasons behind these regulations—that an important factor in their endangerment is the unsustainable use of their parts in traditional medicines." In New York City, however, no shopkeepers mentioned the illegality of these products, much less the conservation status of the species used in their manufacture.

If progress consists of an awareness of the problems caused by using parts of endangered animals, some TCM practitioners in China itself may be moving towards a resolution. In 1995, Beijing's Institute of Materia Medica, Chinese Academy of Medicinal Science produced artificial musk exclusively from synthetic and purified compounds—with no animal components. When tested on one thousand patients as an anti-inflammatory agent, the artificial musk showed the same results as natural musk. Similarly, the Guangdong Provincial Hospital of Traditional Chinese Medicine in China demonstrates a unique combination of TCM and Western medicine and is seen as the leading regional and national authority in medical education, research, treatment, and surgery. The hospital has strict controls regarding the use of animals and now uses cow gall to replace bear bile, buffalo horn to replace rhino horn, and cow bone to replace tiger bone. In a 2003 paper presented at a bear symposium in South Korea, Scarlett Pong, president of the Practicing Pharmacists Association in Hong Kong, stated that many practitioners do not stock or use products from endangered species because, amongst other reasons, "they are morally offensive and there are plenty of perfectly acceptable herbal alternatives."

Perhaps not in New York, but in some Asian countries people who rely on traditional Chinese medicine are sometimes willing to give up the use of animal parts if it can be shown that medications containing them don't work or that the animals providing the parts are in danger of extinction. Because Western medicine often provides cheaper, more efficacious products, Asians in America are much less likely to purchase rhino-horn medications, tiger-bone soup, or bear-balm ointment. These views raise the interesting question of whether various animal-based TCM remedies work. Tiger bone (and other tiger parts) have not been shown to have medicinal value (except perhaps to those who already believe they do); but many people believe that rhino horn, so long a standard in the Chinese medicine pharmacopoeia, actually works to reduce fevers, and they point to some studies to back up their claims.

The primary ingredients of rhino horn are keratin and eukeratin. The horn also contains other proteins, peptides, free amino acids, and cholesterol. Most researchers believe, based on current data, that rhino horn and the horn of the water buffalo have basically similar properties, but in Chinese medicine only rhino horn has been valued for centuries as a medicine used to reduce fevers, calm convulsions, stop nosebleeds, and prevent strokes. Research has demonstrated that using powdered rhino horn is ineffective as a sexual stimulant, and to date there is no conclusive research to demonstrate that the horn has any value when used by traditional herbalists to treat particular life-threatening fevers. After a series of controlled tests, the Swiss pharmaceutical firm Hoffman-La Roche declared that rhino horn has no effect whatsoever on the human body. Scientists in Hong Kong, however, found that rhino horn did have a cooling effect on fever in laboratory rats, but only when used in large doses; humans were not used in the study. Western and TCM practitioners tend to focus on the research results that are supported by their world views; TCM relies upon thousands of years of tradition, whereas Western medicine, for the most part, is a function of replicable scientific research.

At one time, there may have been as many as one hundred thousand wild tigers in India. Because they are largely secretive animals, one could never see large numbers of tigers at any given time, but their occasional propensity to eat livestock—or, more to the point, people—made humans inescapably aware of their presence. In a densely populated country like India, even if only a small percentage turn to man-eating, one hundred thousand tigers is far too many. "At a conservative estimate tigers have consumed well over half a million Indians in the past four centuries," comments James Clarke in *Man Is the Prey*. "In the whole of Asia the figure for the same period cannot be less than a million. Entire districts have been depopulated and villages abandoned, sometimes for years, because of man-eaters." Because tigers have killed so many people, it is not surprising that people would vengefully kill the tigers. As David Quammen put it simply, "Man-eating is the most fatal of indiscretions, in that it often provokes retaliatory eradication."

The world's tiger populations have not primarily been decimated because of their occasional anthropophagous inclinations, of course. Even if tigers, like rhinos, were no threat to humans, I suspect we would have invented reasons to kill them anyway, because their beauty and power makes them one of the world's most desirable "big game animals." We long to believe that in shooting such an animal, these attributes might pass on to us and that in the act of killing we will have surpassed its power and will reign supreme over the animal kingdom.

Indeed, such was the human compulsion to kill tigers that, as we have seen, India's royalty had tigers driven toward their guns as they perched safely atop elephants, and when their primacy was superseded by British occupying forces, the British did likewise, shooting the tigers not only from elephant-back but also on foot, tracking them through the jungle and shooting them from platforms in trees, luring the tiger with a tethered bait animal. The only trophy more spectacular than the orange and black striped skin, often with the snarling, glassy-eyed head attached, was the entire body of the tiger, rearing in a menacing pose, wild-eyed, teeth bared, and claws unsheathed. Man the hunter was so dominant over nature that he could bring a man-eating tiger that he'd rendered harmless into his own living room.

Around the world, the trade in tiger parts continues and not just for TCM. In June 2002, a Rochester, New York, man named Theodore Musson was arrested for selling a snow leopard blanket and a tiger-skin rug over the Internet. Musson also advertised a clouded leopard rug, a jaguar rug, a cheetah head, and a mounted baby tiger for sale. An undercover agent from the U.S. Fish and Wildlife Service arranged to purchase the snow leopard blanket and the tiger-skin rug for $25,000, and Musson was arrested for trafficking in the interstate sale of protected wildlife (McGlynn 2003).

In *The Tiger Skin Trail*, a report compiled by the international Environmental Investigation Agency (EIA) in 2004, Debbie Banks and Julian Newman observe that "while great effort has gone into addressing the demand for tiger bone used in traditional medicines, far less attention has been devoted to the international illegal trade in tiger and leopard skins." Because the trade in skins is dishonest and illegal, there are no easily available records, and the evidence consists largely of contraband shipments seized by customs officers. In August 2003, "the seizure that woke up the world to the bone trade" took place in Delhi: 287 kilograms (630 pounds) of tiger bone, eight tiger skins, and forty-three leopard skins. One of the arrested individuals confessed that the shipment was to be smuggled into China via Ladakh in northern India.

The beautifully marked tiger and leopard skins are made into coats and rugs. In October 2003, the magnitude of the skin trade was revealed when a consignment consisting of 31 tiger skins, 581 leopard skins, and 778 otter skins was found in a truck headed for China from Nepal. A Chinese official was quoted as saying that it was the largest such seizure since 1951. "Though the skin trade is poorly understood and the end markets quite diffuse," wrote Banks and Newman, "it is clear that China is the primary destination for tiger and leopard skins from India. Traders in Tibet have told the EIA that they sell tiger and leopard skins to wealthy Chinese and Europeans, while skin is also used locally as trim on traditional costumes." It is painfully obvious that in the minds of hunters operating in the mountainous hinterlands of the Himalayas, the money that can be earned by poaching and selling endangered cat products far surpasses any puny conservationist ethic.

Thousands of miles from the Himalayas, Australia's not-insignificant population of southern Asian immigrants has made the land down under a mecca for distributors of traditional Chinese medicaments. Since 1999, the Australian government has seized more than twenty-nine thousand illegal imports of TCM products, mostly brought in by travelers for personal use. In August 2004, the customs service announced another seizure of rhino, tiger, and bear products, part of the A$1.5 billion (US$1.15 billion) yearly trade in what Aussies call "complementary medicine." The seizures followed raids on five alternative medicine stores in Brisbane, Sydney, and Melbourne where customs officers and federal police also netted boiled-down monkeys, deer musk, and squashed gecko.

As people occupy more and more of the world's wild areas, they must perforce remove the competition, and no animal was ever more accomplished at competing with man for territory than the tiger. If you place tiger food in front of a hungry tiger, chances are the tiger will take advantage of your generosity and eat the food. People who maintain livestock in areas inhabited by tigers find that the tigers occasionally take a cow, a goat, or a buffalo, and because people cannot tolerate wild animals that threaten their livestock, they feel that they must eliminate the carnivores. Think of the near-total elimination of wolves or the destruction of eagle and mountain lion populations in North America, where there is simply no place for large-prey predators near human settlements. Livestock predation is maddening, but the possibility that humans themselves are potential prey will never be tolerated.

But it is not their beauty, their livestock predation, or even their occasional predation on humans that most threatens tigers today. Rather, their biggest threat is our human desire to transfer the essence of the tiger's strength and virility to our own bodies. Until recently there was little evidence that tigers themselves were hunted for food. But even this practice is apparently on the rise. For instance, a 2003 report published by the Wildlife Protection Society of India described an adult male tiger electrocuted in a trap set by villagers on the outskirts of the Bandhavgarh Tiger Reserve, apparently for use as "bushmeat"—wildlife that is caught and consumed by local inhabitants when their other food sources are exhausted. Examination of the trap suggested that it had been used numerous times in the past, so "bushmeat hunting"—a major problem in central Africa—may have arrived in India as another threat to the tigers. Along with five known instances of tiger electrocution, elephants, rhinos, and leopards have also been killed by poachers in this manner. Further, we learn in *Tiger Skin Trail* (Banks and Newman 2004), a recent report documenting illicit trade in tiger skins, that eating tiger meat is not only for the poor but also for the mainstream and the fashionable: "In late 2003 a series of raids by the Thai Royal Forest Police uncovered a gruesome trade in the meat of captive bred tigers. In one incident in Nonthaburi Province, six live tigers, 22kg of tiger meat and 48kg of tiger bones were recovered from a house along with other specimens of wildlife. Reports suggest that the contraband was destined for restaurants in China, while others suggested the destination was Chinatown in Bangkok." These relatively small-scale trends aside, it is the pharmaceutical demands, tiger essence delivered in the form of pills, tonics, or plasters, that have brought tiger populations to a low from which they may never recover. We are eliminating the tiger in order to manufacture medications of dubious usefulness—even though the pharmacopoeia of Western medicine offers cheaper and readily available substitutes for most of the maladies for which tiger bone is claimed to be useful.

Recent advances in modern medicine might possibly reduce the demand for tiger bone and related substances believed to enhance sexual potency. Viagra is the trade name of a little blue pill used to treat erectile dysfunction in men. Since it first became available in 1998, Viagra (sildenafil citrate) and similar drugs have been recommended by thousands of doctors as therapy for millions of men. Not a hormone or an aphrodisiac, it works by increasing blood flow to the penis. Although the similarity is likely a coincidence, the term *vyaghra* is Sanskrit for "tiger" and thus "Viagra" is a particularly clever name for a medication to cure what is sometimes called "impotence." It can be bought online easily (but not particularly cheaply) and is available in China, South Korea, Taiwan, Singapore, and other countries where tiger bone is the medication of choice for impotence. On the black market, one (genuine) pill can sell for $36. (In the United States, it costs about $8 a tablet.) Compared to the $350 or more that men will pay for a bowl of tiger penis-bone soup, Viagra is a bargain. Because tiger penis-bone soup works no better than fake Viagra tablets, genuine Viagra (or one or another of its competitors) is not only a bargain, but it might actually achieve the desired results for men and at the same time save the lives of tigers. As always, though, the question remains: will users of traditional Chinese medicine forsake millennia of natural cures for the artificial ones of Western medicine? Viagra was approved for use in China in mid-2000, but within six months, counterfeits of the blue pills were appearing in pharmacies around the country. The state-run *China Daily* reported that police arrested two people for running a factory in Shanghai that produced 455,000 counterfeit Viagra tablets.

As with tiger bone, when the real thing is not available, people go to great lengths to fake it. For example, a "substantial amount" of fake tiger parts was brought into Sarawak, Malaysia, from Bangalore, India, and seized by customs officials (Azlan 2004). The haul consisted of 28 fake tiger skins, 1 fake leopard skin, 6 packets of teeth, 56 pairs of "tiger paws," 1,026 pairs of claws, 30 tails, and 45 organ pieces. The skins were painted sheepskins; the teeth and nails were carved from buffalo horn. Malaysia became a signatory to CITES in 1997, which proscribes penalties for the possession or sale of endangered species, but no penalties could be imposed because the "tiger parts" were not from endangered species at all, but from sheep, cattle, and buffalos.

As the world's wild tiger population continued to fall, there were some strange solutions proposed to rescue some of them. In autumn 2002, the Chinese Forestry Administration and the Save China's Tigers Trust of South Africa reached an agreement to send several Chinese tiger cubs to South Africa in August 2003, "where they will undergo a rewilding programme to be reintroduced into the wild. . . . The Chinese tigers that will successfully regain hunting skills and are able to survive independently in the wild will be returned to a Pilot Reserve in China" (*Cat News*, Autumn 2004). A *Cat News* editorial a year later (Autumn 2003) concluded that several IUCN groups, along with a number of scientists and prominent South African conservationist groups, had many reservations about the plan: "They do not think that it is the most effective plan for the conservation of the tiger in China. We further submit that the release of free-ranging tigers in South Africa is not in the best interests of biodiversity conservation in South Africa."

In another misguided effort to "rescue" endangered wild tigers, South African filmmaker John Varty and Dave Salmoni, a Canadian zoologist, took two young tigers born in a Cincinnati zoo and brought them to South Africa, with the idea of introducing them into the "wild." That tigers are not now and never have been in Africa did not deter them; their plan was to teach the pair of tigers—a brother and sister named Ron and Julie—to hunt wild prey in a large fenced enclosure, in preparation for releasing them into a larger enclosure where they would hunt and kill on their own. From 1999 to 2002, John filmed Dave's efforts on behalf of "tiger conservation," but the whole enterprise looked like nothing more than a stunt. The full-page ad announcing this Discovery Channel special on September 14, 2003, blared, "Stalked to the Edge of Extinction. There's One Last Hope." Tigers are indeed scarce in India—the current best guess is that there are 3,500—but introducing zoo-raised tigers that have been trained to hunt wildebeest in Africa will do nothing for the depleted Indian tiger population.

To its credit, the Discovery Channel asked various tiger experts, including Dr. John Seidensticker, a senior scientist at the Smithsonian's National Zoological Park, to comment on the program. The responses were posted on the Discovery Channel's Web site. Seidensticker succinctly summed up his opposition to the project: "I've heard many justifications for this project. . . . but the conservation community is pretty much opposed to it. . . . Our definition of conservation is securing a place for wild tigers where they live, not a place in Texas or South Africa. There are a lot of people who spent their lives, sometimes at great risk to themselves, to work on tiger conservation. It's going to be a story, this whole thing, about how *not* to do conservation."

"The tiger," wrote IUCN cat specialist Peter Jackson in 1997, "has always been regarded by humans with awe. Admiration for its beauty has been combined with underlying fear of its massive power." Some perversion of this attitude has led to the bizarre idea of keeping these giant, dangerous animals as pets. Celebrities Michael Jackson and Mike Tyson each have one; Siegfried and Roy have an actful in Las Vegas; Internet and specialty trade magazines advertise exotic-animal auctions and "jungle-cat reduction sales." The prices are not particularly prohibitive: $500 for a tiger cub; $5,000 for a pair of Bengal tigers; up to $15,000 for a more fashionable white tiger. Tigers, which adapt well to captivity, are also well represented in zoos because they are among nature's most glorious creatures and people like to look at them. Today in Texas there are said to be four thousand pet tigers, more perhaps than the number that roam free in India, and because captive tigers are just as fertile as domestic cats, the numbers are likely to grow. There are said to be four hundred to five hundred lions and tigers in the Houston area alone (Handwerk 2003).

On October 3, 2003, one of the white tigers used in the Las Vegas act of Siegfried and Roy attacked Roy by grabbing his arm, and when the tiger wouldn't release him, Roy hit him on the head with the microphone. This evidently angered the tiger to the point where it bit Roy on the neck and nearly killed him, picking him up and dragging him off, just as it would do with a prey animal. On October 30, 2003, Roy was taken off the critical list in a Las Vegas hospital; the tiger was in isolation, and Siegfried and Roy's shows were cancelled indefinitely. The following day, there was another story about a tiger, this one that lived in a New York City apartment. It seems that Mr. Antoine Yates had a 400-pound tiger named "Ming" in his fifth-floor apartment in Harlem, and some of the neighbors were complaining about the smell. Ming apparently had become dangerous, even to Yates, and he moved into an adjacent apartment and fed the tiger by throwing raw chickens through the narrowly opened door (Polgreen and George 2003). The police went to Yates's apartment, drilled a small hole through the door, and were more than a little surprised at the sight of a full-grown tiger crouching inside. The tiger was shot through the window with a tranquilizing dart and removed from the apartment.

The attack on Roy, coming as it did almost simultaneously with the discovery of Yates's home-raised tiger, suddenly raised America's consciousness about tigers, at least for the moment. New York tabloids splashed dramatic photographs on their front pages (remember, tigers

SUNDAY

NEW YORK POST 50¢

LATE CITY FINAL

SIEGFRIED & ROY HORROR

TRAGIC TRICK

Beast puts bite on old pal

Campy magician Roy Horn (right) of Siegfried & Roy is fighting for his life in a Las Vegas hospital after being mauled by a white male Siberian tiger he's owned for six years.

Last night, 40 minutes into a routine they've done hundreds of times, the big cat lunged at Horn's neck and dragged him off stage like prey.

STORY: PAGE 7

YANKS 1 WIN AWAY

SEE SPORTS

Tigers are the most popular animals in the world, but sometimes things go wrong. In October 2003 in Las Vegas, one of Siegfried and Roy's white tigers attacked Roy and nearly killed him.

sell), and almost everybody had something to say about tigers in show business, tigers in the home, and even tigers in the wild. Two weeks after the tiger news broke, novelist Charles Siebert wrote this in the *New York Times Magazine*:

> "We can judge the heart of man," wrote Immanuel Kant, "by his treatment of animals." It may be, in the end, fear above all other emotions that moves us to want to kidnap them and keep them close by, living emissaries of a primal world and self that we've long left behind. They are, in a sense, all that we have left of that world, just as we are becoming their only keepers in this one.

Rhinos don't make very good pets, nor are they particularly dangerous, so we must look elsewhere for the reason for their impending extinction:

in nearly all cases, they are killed solely for their horns, a staple of TCM. In the not-so-distant past, people like Ernest Hemingway hunted them for the thrill of shooting a very large animal with a very powerful gun from a very great distance. Their taxidermied heads were hung on the wall and their feet were turned into umbrella stands, but overall demand for such trophies was modest. In reality, the majority of rhinos dying at the hand of *Homo sapiens* during the past century were killed to meet medicinal demand for their horns, with a substantial number being killed for jambiya handles. The three Asian rhino species—Indian, Javan, and Sumatran—were the first to go, hunted almost to extinction, but the horn of the African black rhino, highly prized for carving dagger handles, is also actively sought. The seemingly innocuous pursuit of carving dagger handles has brought the black rhino to such low levels that its very existence is threatened. A century ago, there were an estimated 1 million black rhinos in Africa. There are now perhaps 3,600, and the numbers are falling. White rhinos, Africa's other representative of the nose-horned ones, have also been hunted for medicine and jambiyas, but, more tractable than their somewhat smaller "black" cousins, they have been successfully transplanted to reserves where they may survive unmolested.

Unlike tiger bone or rhino horn, bear bile can be legally prescribed in many countries, and, despite the CITES listing of the moon bear as an endangered species, bear products from gall bladders to paws still appear in pharmacies and on menus. It now appears that Chinese practitioners were not altogether wrong about the usefulness of bile acid and even underestimated the variety of conditions it could treat.

With his colleagues at the University of Minnesota's departments of medicine and neurosurgery, Clifford Steer has discovered that bile acid (ursodeoxycholic acid) can reduce brain damage by more than 50 percent in stroke-impaired laboratory rats, for instance (Rodrigues et al. 2002). Steer discovered that bile acid is a potent antiapoptotic agent (apoptosis is the death of cells). Specifically, ursodeoxycholic acid (from *urso* = "bear") has been used as a therapeutic agent to treat models of Huntington's disease, head trauma, and acute stroke, as well as Parkinson's disease (Keene et al. 2002). One common characteristic of these disorders as well as some others is the role that apoptosis plays, as found on Steer's University of Minnesota faculty Web page:

> Bile acid has been determined as a potent antiapoptotic agent, significantly improving neurologic status in these models. In the basic science studies, the lab has delineated the molecular mechanism by which ursodeoxycholic acid acts to preserve cell survival and cell function. As a therapeutic agent, ursodeoxycholic acid is unique in that it is a natural bile acid with no significant toxicity, crosses the blood-brain barrier, and can be delivered easily to patients. There are, in fact, many disease states that could potentially benefit, including myocardial infarction, autoimmune diseases, and the many acute and chronic neurodegenerative disorders for which there is little available treatment.

After a stroke had been mechanically induced in rats, researchers injected some of the rats with ursodeoxycholic acid and some with a neutral substance. When the rats were tested two days later, those that had been injected with ursodeoxycholic acid were found to have suffered considerably less neurological damage than those in the control group. In human stroke, a 50 percent destruction of brain cells almost completely disables the victim, and "a 10–20 percent reduction in damage could be the difference between a patient walking out of the hospital or being pushed out in a wheelchair," Steer said in a March 2002 interview with the *St. Paul Pioneer Press*, a Minnesota newspaper (Majeski 2002). For his research, Steer orders ursodeoxycholic acid from American chemical supply houses. When he queried several of the commercial vendors as to where they had obtained the substance, he was told that the information regarding sources is proprietary. When I spoke to Steer in February of 2004, I asked if it was possible that the American chemical supply houses were getting bear bile from Chinese or other Asian bear farms. He said he didn't know and they wouldn't tell him. But in a 2005 article in *Wildlife Tracks*, the newsletter of the Humane Society of the United States, Adam Roberts, Executive Director of the Animal Welfare Institute, noted that a sting operation in Virginia had "uncovered evidence that whole bears, bear gallbladders, bear paws, and other bear parts originating in Virginia are being trafficked to Washington, D.C., North Carolina, New Jersey, New York, and California, as well as overseas." He concluded that "on the black market, bear parts, particularly the gallbladder and bile, literally are worth their weight in gold, and can fetch more than gems or drugs. The illicit international trade in bear parts not only puts endangered Asiatic bear species in further danger, but it has put a price on the head of every black bear in America."

It is possible, then, that while a program is under way to protect and free the bears on Chinese farms, the gall bladders are becoming even more valuable because of the recent discovery that bear bile, which is substantially richer in ursodeoxycholic acid than cow or pig bile, does actually work to ameliorate some diseases in humans. (Steer's work has so far been conducted mostly on laboratory rats and mice, but he indicated that human clinical trials are scheduled to begin in mid-2005 for certain neurodegenerative diseases.) It is conceivable, therefore, that even as synthetic bear bile is being developed, bears will continue to be killed for their gall bladders by unscrupulous collectors.

Too many animals, from sea horses to rhinoceroses, are endangered by the demands of traditional Chinese medicine. Of course, TCM is not the only factor in the endangerment of these animals, but it plays an enormous part. If present trends continue, tigers and rhinos will become extinct in the wild, perhaps in our lifetime and almost certainly in the lifetime of our children's children. Our offspring will know what a tiger is—we live in an age of eternally preserved and eternally available video images—and captive tigers in homes, cages, zoos, and circuses will ensure that these powerful cats will not disappear entirely, like the dodo or the passenger pigeon. For the time being, it is enough to know that the tiger still prowls the dark jungles of Asia and the darker jungles of our consciousness. But when the last wild tiger is gone, a bit of every one of us will go with it not because it is big, beautiful, or dangerous, but because it awakens in us the sense of what it means to be a human being, alive in the world at this time. Invoking the sort of mysticism that does not involve the death of animals, but rather its opposite, author Henry Beston wrote, "We need another and a wiser and perhaps a more mystical concept of animals. Remote from universal nature, and living by complicated artifice, man in civilization surveys the creature through the glass of his knowledge and sees thereby a feather magnified and the whole image in distortion."

Surely the most distorted image we can conjure is that animals were put here for our use. Domestication for food is easy to rationalize—it will be a long time before people stop eating cows, pigs, or chickens (not to mention dogs, banded civets, pangolins, or snakes)—and the use of oxen, camels, and water buffalos as beasts of burden is not likely to cease until third-world agriculturalists obtain inexpensive engines. Because we shared our early caves and campfires with proto-dogs, we will probably continue to do so. But enlightened humans (*Homo sapiens*, remember)—even those who have used animal parts for medicine for thousands of years—should recognize that the animals that provide these pharmaceuticals, whether sometimes useful or not, are becoming extinct. This equation is painfully obvious: No more tigers or rhinos means no more tiger bone or rhino horn. It is critically important to develop substitutes for animal substances and just as important to develop a heightened awareness of the precarious status of some of the hunted animals. Beston's quote concludes thus: "For the animal shall not be measured by man. In a world older and more complex than ours they move finished and complete, gifted with extensions of the senses we have lost or never attained, living by voices we shall never hear. They are not brethren, they are not underlings; they are other nations, caught with ourselves in the net of life and time, fellow prisoners of the splendour and travail of the earth."

Bibliography

Abbot, N. C., A. R. White, and E. Ernst. 1996. Complementary medicine. *Nature* 381:361.

Agusti, J., and M. Anton. 2002. *Mammoths, Sabertooths, and Hominids: 65 Million Years of Mammalian Evolution in Europe*. Columbia University Press.

Ale, S. 2000. Conservation of the snow leopard in Nepal. *Cat News* 32:8–9.

Allen, J. A. 1898. Fur seal hunting in the Southern Hemisphere. Pp. 307–319 in *The Fur Seals and the Fur-Seal Islands of the North Pacific Ocean*, Vol. 3, edited by D. S. Jordan. Government Printing Office.

Animals Asia. 2003. http://www.bearrescue.org/aboutAAF.html.

Anon. 2000. Snow leopard smuggler arrested in China. *Cat News* 32:12.

———. 2003a. Big cat uses infrasound to keep rivals at bay. *New Scientist* 178 (2393): 21.

———. 2003b. Woman helps bears claw their way to freedom. *National Enquirer*, September 20.

———. 2004a. Birds attack cats. *Science* 306:808.

———. 2004b. India's most famous tiger reserve threatened by tourism. *Cat News* 40:4.

———. 2005c. No rhino rescue. *New Scientist* 185(2486): 7.

Aristotle, N. D. *Historia Animalium*. Loeb Classical Library, Harvard University.

Ashby, K. R., and C. Santiapillai. 1987. An outline strategy for the conservation of the tiger (*Panthera tigris*) in Indonesia. Pp. 411–415 in *Tigers of the World: The Biology, Biopolitics, Management, and Conservation of an Endangered Species*, edited by R. L. Tilson and U. S. Seal. Noyes.

Ashton, J. 1890. *Curious Creatures in Zoology*. Cassell.

Avasthi, A. 2004. Plant mimic may be cure for malaria. *New Scientist* 183 (2461): 15.

Azlan, M. J. 2004. Fake tiger parts seized in Sarawak. *Cat News* 40:6–7.

Baillio, W. 1994. *The Wild Kingdom of Antoine-Louis Barye, 1795–1875*. Wildenstein.

Banks, D., and J. Newman. 2004. *The Tiger Skin Trail*. Environmental Investigation Agency.

Baoagang, S., E. Zhang, and D. Miquelle. 1999. Siberian tigers on brink of extinction in China. *Cat News* 31:2.

Baratta, K. 2003. Decision could clear way for tigers' move. *Allentown (NJ) Examiner*, May 15.

Barnett, A. 2003. West's love of talc threatens Indian tigers. *Guardian Weekly*, June 26–July 2.

Bartlett, D., and J. Bartlett. 1992. Africa's skeleton coast. *National Geographic* 181 (1): 54–85.

Bibliography

Beard, P. H. 1965. *The End of the Game*. Viking.

Beer, R. R. 1977. *Unicorn: Myth and Reality*. Mason/Charter.

Beigbeder, O. 1965. *Ivory*. G. P. Putnam's Sons.

Bensky, D., and A. Gamble, trans. 1993. *Chinese Herbal Medicine: Materia Medica*. Eastland Press.

Berger, J. 1993. Rhino conservation tactics. *Nature* 361:121.

———. 1994. Science, conservation, and black rhinos. *Journal of Mammalogy*. 75:298–308.

Berger, J., and C. Cunningham. 1994a. Active intervention and conservation: Africa's pachyderm problem. *Science* 263:1241–1242.

———. 1994b. Phenotypic alterations, evolutionary significant structures, and rhino conservation. *Conservation Biology* 8:833–840.

———. 1995. Predation, sensitivity, and sex: Why female black rhinoceroses outlive males. *Behavioral Ecology* 9:57–64.

———. 1996. Is rhino dehorning scientifically prudent? *Pachyderm* 21:60–68.

Berger, J., C. Cunningham, and A. A. Gasuweb. 1994. The uncertainty of data and dehorning black rhinos. *Conservation Biology* 8 (4): 1149–1152.

Berger, J., C. Cunningham, A. A. Gasuweb, and M. Lindeque. 1993. "Costs" and short-term survivorship of hornless black rhinos. *Conservation Biology* 7 (4): 920–924.

Bertram, B. 1997. Review of *On the horns of a dilemma*. *Nature* 388:636.

Beston, H. 1928. *The Outermost House*. Holt, Rinehart and Winston.

Bonner, R. 1993. *At the Hand of Man: Peril and Hope for Africa's Wildlife*. Knopf.

Bonner, W. N. 1982. *Seals and Men: A Study of Interactions*. Washington Sea Grant.

Boomgaard, P. 2001. *Frontiers of Fear: Tigers and People in the Malay World, 1600–1950*. Yale University Press.

Borner, M. 1978. Status and conservation of the Sumatran tiger. *Carnivore* 1 (1): 97–102.

Bosi, E. J. 1996. Mating Sumatran rhinoceros at Sepilok Rhino Breeding Centre. *Pachyderm* 21:24–27.

Bradsher, K. 2004. China to kill 10,000 civet cats in effort to eradicate SARS. *New York Times*, January 5.

Bradsher, K., and L. K. Altman. 2003. Strain of SARS is found in 3 animal species in Asia. *New York Times*, May 23.

Breeden, S. 1984. Tiger! Lord of the Indian jungle. *National Geographic* 166 (6): 748–773.

Breeden, S., and B. Wright. 1996. *Through the Tiger's Eyes*. Ten Speed Press.

Breur, M. 2003. Las Vegas stage fright: Roy struck tiger before it snapped. *New York Post*, October 5.

Brooke, J. 2005. Siberian tiger hunt seeks to save, not shoot, the big cats. *New York Times*, January 16.

Browne, A. 2002. Store's trade in rare leopard bones exposed. *Observer*, May 12.

Bruemmer, F. 1977. *The Life of the Harp Seal*. Times Books.

———. 1993. *The Narwhal: Unicorn of the Sea*. Key Porter.

Burger, J. 1991. A passage to India [Kaziranga]. *Wildlife Conservation* 84 (6): 72–77.

Bibliography

Burn-Murdoch, W. G. 1917. *Modern Whaling & Bear-Hunting*. Seeley, Service.

Burton, R. G. 1936. *The Tiger Hunters*. London. (Mittal edition, New Delhi, 2002.)

Busch, B. C. 1985. *The War Against the Seals: A History of the North American Seal Fishery*. McGill-Queens University Press.

But, P. P.-H., L.-C. Lung, and Y.-K. Tam. 1990. Ethnopharmacology of rhinoceros horn. *Journal of Ethnopharmacology* 30:157–168.

Byrne, M. St.C. 1926. *Elizabethan Zoo*. Haslewood Books.

Capstick, P. H. 1981. *Maneaters*. Safari Press.

Carter, R. 1996. Holistic hazards—millions of people in Britain turn to complementary medicine in the belief that it is risk-free. But a closer look shows that side effects are widespread. *New Scientist* 151 (2038): 12–13.

Cavallo, A. S. 1998. *The Unicorn Tapestries at the Metropolitan Museum of Art*. Abrams.

Chan, S., A. V. Maksimuk, L. V. Zhirnov, and S. V. Nash. 1995. *From Steppe to Store: The Trade in Saiga Antelope Horn*. TRAFFIC International.

Chen, Y.-U., S. Wu, W. H. Bhiksu, and V. Fisher. 2002. Bear markets: Taiwan. Pp. 213–228 in *The Bear Bile Business: The Global Trade in Bear Products from China to Asia and Beyond*, edited by T. Phillips and P. Wilson. World Society for the Protection of Animals.

Chestin, I., and V. Yudin. 1999. Status and management of the Asiatic black bear in Russia. Pp. 211–213 in *Bears—Status Survey and Action Plan*, compiled by C. Servheen, S. Herrero, and B. Peyton. IUCN, Gland, Switzerland.

Choudhury, A. 1997. The status of the Sumatran rhinoceros in north-eastern India. *Oryx* 31 (2): 151–152.

———. 1998. Flood havoc in Kaziranga. *Pachyderm* 26:83–87.

Chundawat, R. S., N. Gogate, and A. J. T. Johnsingh. 1999. Tigers in Panna: Preliminary results from an Indian tropical dry forest. Pp. 123–129 in *Riding the Tiger: Tiger Conservation in Human-Dominated Landscapes*, edited by J. Seidensticker, S. Christie, and P. Jackson. Cambridge University Press.

Clarke, J. 1969. *Man Is the Prey*. Andre Deutsch.

Clarke, T. H. 1986. *The Rhinoceros from Dürer to Stubbs—1515–1799*. Sotheby's.

Coghlan, A. 2003a. Chinese herb hits malaria where it hurts. *New Scientist* 179:16.

———. 2003b. Forensic test fingers rhino poachers. *New Scientist* 179 (2411): 9.

Conrad, K. 2000. Safety in numbers: Review of the breeding center for Felidae at Hengdaohezi. IUCN/SSC Cat Specialist Group. http://www.5tigers.org/Research/southchina/China/ChineseCenter.pdf.

Corbett, J. 1946. *Man-Eaters of Kumaon*. Oxford University Press.

———. 1954. *The Temple Tiger and More Man-Eaters of Kumaon*. Oxford University Press.

Culpeper, N. 1653. *The Complete Herbal and English Physician Enlarged*. London. (1995 edition, Wordsworth.)

Cunningham, C., and J. Berger. 1997. *Horn of Darkness: Rhinos on the Edge*. Oxford University Press.

Cyranoski, D. 2004. Campaign to fight malaria hit by surge in demand for medicine. *Nature* 432:259.

Bibliography

Dance, S. P. 1978. *The Art of Natural History*. Overlook Press.

Dandurant, K. 2003. Seal deaths investigated. *Portsmouth (NH) Herald*, October 30.

Daniel, J. C. 2001. *The Tiger in India: A Natural History*. Natraj.

Dexel, B. 2002. *The Illegal Trade in Snow Leopards—A Global Perspective*. German Society for Nature Conservation.

Diamond, D. 2004. Murder most beastly. *National Geographic* 205 (3).

Dinerstein, E. 2003. *The Return of the Unicorns: The Natural History and Conservation of the Greater One-Horned Rhinoceros*. Columbia University Press.

Dinerstein, E., A. Rijal, M. Bookbinder, B. Kattel, and A. Rajuria. 1999. Tigers as neighbors: Efforts to promote local guardianship of endangered species in lowland Nepal. Pp. 316–333 in *Riding the Tiger: Tiger Conservation in Human-Dominated Landscapes*, edited by J. Seidensticker, S. Christie, and P. Jackson. Cambridge University Press.

Dixon, C. 2004. Last 39 tigers are moved from unsafe rescue center. *New York Times*, June 11.

Doherty, J. 2002. Tigers at the gate. *Smithsonian* 32 (10): 66–67.

Dugmore, A. R. 1925. *The Wonderland of Big Game: Being an Account of Two Trips Through Tanganyika and Kenya*. Arrowsmith.

du Toit, R. 2002. Black rhino crisis in Zimbabwe. *Pachyderm* 32:83–85.

Eckstein-Ludwig, U., R. J. Webb, I. D. A. van Goethem, J. M. East, A. G. Lee, M. Kimura, P. M. O'Neill, P. G. Gray, S. A. Ward, and S. Krishna. 2003. Artemesinins target the SERCA of *Plasmodium falciparum*. *Nature* 424:957–961.

Eisenberg, D. M., R. B. Davis, S. L. Ettner, S. Appel, S. Wilkey, M. Van Rompay, and R. C. Kessler. 1998. Trends in alternative medicine use in the United States 1990–1997: Results of a follow-up national survey. *Journal of the American Medical Association* 280:1569–1575.

Eisenberg, D. M., R. C. Kessler, C. Foster, F. E. Norlock, D. R. Calkins, and T. L. Delbanco. 1993. Unconventional medicine in the United States—prevalence, costs, and patterns of use. *New England Journal of Medicine* 328:246–252.

Emslie, R. 2000. African rhinos numbering 13,000 for the first time since the mid-1980s. *Pachyderm* 29:53–56.

———. 2002. African rhino numbers continue to increase. *Pachyderm* 33:103–107.

Environmental News Service (ENS). 2004. 30 Thai tigers die of bird flu; 30 more to be culled. http://www.ens-newswire.com/login/index.asp?q=/ens/oct2004/2004-10-21-04.asp.

Erlande-Brandenburg, A. 1983. *La Dame à la Licorne*. Editions de la Réunion des Musées Nationaux.

Ewald, P. W. 1994. *Evolution of Infectious Disease*. Oxford University Press.

———. 2002. *Plague Time: The New Germ Theory of Disease*. Random House.

Ewald, P. W., and G. Cochran. 1999. Catching on to what's catching. *Natural History* 108 (1): 34–37.

Fernando, P., T. N. C. Vidya, J. Payne, M. Stuewe, G. Davison, R. J. Alfred, P. Andau, E. Bosi, A. Kilbourn, and D. J. Melnick. 2003. DNA analysis indicates that Asian elephants are native to Borneo and are therefore a high priority for conservation. *Public Library of Science Biology* 1 (1): 1–6.

Fertl, D., M. Reddy, and E. D. Stoops. 2000. *Bears*. Sterling.

Feuer, A., and J. George. 2003. Police subdue tiger in Harlem apartment. *New York Times*, October 5.

Finley, K. J., R. A. Davis, and H. B. Silverman. 1980. Aspects of the narwhal hunt in the eastern Canadian Arctic. *Reports of the International Whaling Community* 30 (SC/31/SM9) 459–464.

Fisher, J., N. Simon, and J. Vincent. 1969. *Wildlife in Danger*. Viking.

Foose, T. J., and N. J. van Strien. 1998. Conservation programmes for Sumatran and Javan rhinos in Indonesia and Malaysia. *Pachyderm* 26:100–115.

Foster, S. J., and A. C. J. Vincent. 2004. Life history and ecology of seahorses: Implications for ecology and management. *Journal of Fish Biology* 65 (1): 1–61.

Fox, J. L., and Du Jizeng, eds. 1994. *Proceedings of the Seventh Annual Snow Leopard Symposium*. International Snow Leopard Trust and Chicago Zoological Society.

Franklin, N., S. Bastoni, D. Siswomartono, J. Manasang, and R. Tilson. 1999. Last of the Indonesian tigers: A cause for optimism. Pp. 130–147 in *Riding the Tiger: Tiger Conservation in Human-Dominated Landscapes*, edited by J. Seidensticker, S. Christie, and P. Jackson. Cambridge University Press.

Freeman, H., ed. 1988. *Proceedings of the Fifth Annual Snow Leopard Symposium*. International Snow Leopard Trust and Wildlife Institute of India.

Freeman, M. B. 1976. *The Unicorn Tapestries*. Metropolitan Museum of Art.

Friel, D. 2005. Indian PM orders moves to save disappearing tigers. *Reuters Report* http://reuters.com/newsArticle.jhtml?type=topNews&storyID=794744.

Fruehauf, H. 1999. Chinese medicine in crisis. *Journal of Chinese Medicine* 61:1–9.

Gahaku, C. G. 1991. African rhinoceroses: Challenges continue in the 1990s. *Pachyderm* 14:44–47.

Galster, S. R., and K. V. Eliot. 1999. Roaring back: Anti-poaching strategies for the Russian Far East and the comeback of the Amur tiger. Pp. 230–241 in *Riding the Tiger: Tiger Conservation in Human-Dominated Landscapes*, edited by J. Seidensticker, S. Christie, and P. Jackson. Cambridge University Press.

Gee, E. P. 1952. The great Indian one-horned rhinoceros. *Oryx* 1 (5): 224–227.

Gerard, J. 1597. *Historie of Plants*. London. (Reprinted as *Gerard's Herbal*, 1998. Senate.)

Geist, V. 1998. *Deer of the World*. Stackpole.

Gesner, C. 1551–87. *Historia Animalium*. Zurich.

Gies, J., and F. Gies. 1969. *Life in a Medieval City*. Harper Colophon.

Goldsmid, E. 1886. *Un-Natural History, or Myths of Ancient Science; Being a Collection of Curious Tracts on the Basilisk, Unicorn, Phoenix, Behemoth or Leviathan, Dragon, Giant Spider, Tarantula, Chameleons, Satyrs, Homines Caudati, &c*. Edinburgh.

Goodrich, J. 2004a. Against all odds. *Wildlife Conservation* 107 (3): 6–7.

———. 2004b. Team tiger. *Wildlife Conservation* 107 (3): 34–37.

Goodrich, J., D. Miquelle, L. Kerley, and E. Smirnov. 2002. Time for tigers: Paving the way for tiger conservation in Russia. *Wildlife Conservaton* 105 (1): 22–29.

Gould, C. 1886. *Mythical Monsters*. London. (1989 edition, Crescent Books.)

Govind, V., and S. Ho. 2001. *Consumer Report on the Trade in Bear Gall Bladder and Bear Bile Products in Singapore*. Animal Concerns Research and Education Society (ACRES).

Govind, V., S. Ho, G. Subramaniam, B. Sim, and M. Tan. 2002. Bear markets: Singapore. Pp. 185–211 in *The Bear Bile Business: The Global Trade in Bear Products from China to Asia and Beyond*, edited by T. Phillips and P. Wilson. World Society for the Protection of Animals.

Graham-Rowe, D. 2004. Kit to beat bear trade. *New Scientist* 184 (2476): 16.

Grove, N. 1981. Wild cargo: The business of smuggling animals. *National Geographic* 159 (3): 287–316.

Groves, C. P. 1972. *Ceratotherium simum* [white rhinoceros]. *Mammalian Species* 8:1–6. American Society of Mammalogists.

Groves, C. P., and F. Kurt. 1972. *Dicerorhinus sumatrensis* [Sumatran rhinoceros]. *Mammalian Species* 21:1–6. American Society of Mammalogists.

Guggisberg, C. A. W. 1966. *S.O.S. Rhino*. October House.

———. 1966. *Wild Cats of the World*. David & Charles.

Haberman, Z., L. Celona, and L. Greene. 2003. 400-lb cat and gator found in Harlem apt. *New York Post*, October 5.

Hamer, M. 1992. Poachers kill tigers for their bones. *New Scientist* 135 (1829): 5.

Handwerk, B. 2003. Big cats kept as pets across U.S., despite risk. http://news.nationalgeographic.com/news/2002/08/0816_020816_EXPLcats.html.

Harris, J. M., ed. 1992. *Rancho La Brea: Death Trap and Treasure Trove*. Los Angeles Museum of Natural History.

Harrison, J. 1974. *An Introduction to the Mammals of Singapore and Malaysia*. Malayan Nature Society.

Hathaway, N. 1980. *The Unicorn*. Viking.

Hay, K. A., and A. W. Mansfield. 1989. Narwhal—*Monodon monoceros* Linnaeus 1758. Pp. 145–176 in *Handbook of Marine Mammals, Vol. 4, River Dolphins and the Larger Toothed Whales*, edited by S. H. Ridgway and R. J. Harrison. Academic Press.

Hazumi, T. 1999. Status and management of the Asiatic black bear in Japan. Pp. 207–211 in *Bears—Status Survey and Action Plan*, compiled by C. Servheen, S. Herrero, and B. Peyton. IUCN, Gland, Switzerland.

Hemingway, E. 1935. *Green Hills of Africa*. Scribners. (Touchstone edition, 1996.)

Hemley, G., and J. A. Mills. 1999. The beginning of the end of tigers in trade? Pp. 217–229 in *Riding the Tiger: Tiger Conservation in Human-Dominated Landscapes*, edited by J. Seidensticker, S. Christie, and P. Jackson. Cambridge University Press.

Hemmer, H. 1972. *Uncia uncia* [snow leopard] *Mammalian Species* 20:1–5. American Society of Mammalogists.

Henry L. 2004. *A Tale of Two Cities: A Comparative Study of Traditional Chinese Medicine Markets in San Francisco and New York City*. TRAFFIC North America.

Herklots, G. A. C. 1937. The pangolin or scaly ant-eater. *Hong Kong Naturalist* 8 (2):78–83.

Highley, K. 1993. The market for tiger products in Taiwan: A survey. Earthtrust report. http://www.earthtrust.org/tiger.html.

Highley, K., and S. C. Highley. 1994. Bear farming and trade in China. Earthtrust report. http://www.earthtrust.org/bear.html.

Hillman-Smith, A. K. K. 1998. The current status of the northern white rhino in Garamba. *Pachyderm* 25:104–105.

Hillman-Smith, A. K. K., and C. P. Groves. 1994. *Diceros bicornis* [black rhinoceros]. *Mammalian Species* 455:1–8. American Society of Mammalogists.

Hilton-Taylor, C., comp. 2000. *2000 IUCN Red List of Threatened Species*. IUCN, Gland, Switzerland, and Cambridge, UK.

Holland, P., trans. 1601. *The Historie of the World*. C. Plinius Secundus (Pliny the Younger). Adam Islip, London.

Homes, V. 1999. *On the Scent: The Uses of Musk and Europe's Role in Its Trade*. TRAFFIC Europe.

Homes, V., ed. 2004. *No License to Kill: The Population and Harvest of Musk Deer and Trade in Musk in the Russian Federation and in Mongolia*. TRAFFIC Europe.

Hoogerwerf, A. 1970. *Udjung Kulon: The Land of the Last Javan Rhinoceros*. Brill.

Hooper, J. 1999. A new germ theory. *Atlantic Monthly* 283 (2):41–51.

Horn, J. S. 1969. *Away with All Pests: An English Surgeon in People's China: 1954–1969*. Modern Reader.

Hornocker, M. 1997. *Track of the Tiger*. Sierra Club.

Hose, C. 1893. *A Descriptive Account of the Mammals of Borneo*. London.

———. 1929. *The Field Book of a Jungle-Wallah: Being a Description of Shore, River, and Forest Life in Sarawak*. London.

Hu, H., and Z. Jiang. 2002. Trial release of Père David's deer, *Elaphurus davidianus* in the Dafeng Reserve, China. *Oryx* 36 (2): 196–199.

Hunter, J. A. 1952. *Hunter*. Harper & Brothers.

Hussain, S. 2003. The status of the snow leopard in Pakistan and its conflict with local farmers. *Oryx* 37 (1): 26–33.

Huxley, J. 1963. *Wild Lives of Africa*. Harper & Row.

Ives, R. 1996. *Of Tigers and Men*. Doubleday.

Jackson, P. 1990. *Tigers*. Chartwell.

———. 1997. Status of the tiger *Panthera tigris* (Linnaeus 1758) in 1997. *Cat News* 27:9.

———. 1999. The tiger in human consciousness and its significance in crafting solutions for tiger conservation. Pp. 50–54 in *Riding the Tiger: Tiger Conservation in Human-Dominated Landscapes*, edited by J. Seidensticker, S. Christie, and P. Jackson. Cambridge University Press.

Jackson, R., and D. Hillard. 1986. Tracking the elusive snow leopard. *National Geographic* 169 (6): 793–809.

Juliano, A. L., and J. A. Lerner. 2001. *Monks and Merchants: Silk Road Treasures from Ancient China*. Harry Abrams/Asia Society.

Kang, S., and M. Phipps. 2003. *A Question of Attitude: South Korea's Traditional Medicine Practitioners and Wildlife Conservation*. TRAFFIC East Asia.

Karanth, K. U. 1987. Tigers in India: A critical review of field censuses. Pp. 118–132 in *Tigers of the World: The Biology, Biopolitics, Management, and Conservation of an Endangered Species*, edited by R. L. Tilson and U. S. Seal. Noyes.

———. 1999. Counting tigers, with confidence. Pp. 104–105 in *Riding the Tiger: Tiger Conservation in Human-Dominated Landscapes*, edited by J. Seidensticker, S. Christie, and P. Jackson. Cambridge University Press.

———. 2001. *The Way of the Tiger*. Voyageur.

Karanth, K. U., and M. D. Madhusudan. 1997. Avoiding paper tigers and saving real tigers: Response to Saberwal. *Conservation Biology* 11 (3): 818–820.

Karanth, K. U., J. D. Nichols, J. Seidensticker, E. Dinerstein, J. L. D. Smith, C. McDougal, A. J. T. Johnsingh, R. S. Chundawat, and V. Thapar. 2003. Science deficiency in conservation practice: The monitoring of tiger populations in India. *Animal Conservation* 6 (2): 141–146.

Karanth, K. U., and B. M. Stith. 1999. Prey depletion as as a critical determinant of tiger population viability. Pp. 100–113 in *Riding the Tiger: Tiger Conservation in Human-Dominated Landscapes*, edited by J. Seidensticker, S. Christie, and P. Jackson. Cambridge University Press.

Karanth, K.U., M. Sunquist, and K. M. Chinappa. 1999. Long-term monitoring of tigers: Lessons from Nagarahole. Pp. 350–353 in *Riding the Tiger: Tiger Conservation in Human-Dominated Landscapes*, edited by J. Seidensticker, S. Christie, and P. Jackson. Cambridge University Press.

Keay, J. 2000. *India: A History*. Grove Press.

Keene, C. D., C. M. P. Rodrigues, T. Eich, M. S. Chabbra, C. J. Steer, and W. C. Low. 2002. Tauroursodeoxycholic acid, a bile acid, is neuroprotective in a transgenic animal model of Huntington's disease. *Proceedings of the National Academy of Sciences*. 99:10671–10676.

Kellogg, D. 1989. *In Search of China*. University of Hawaii Press.

Kemf, E., and N. van Strein. 2002. *Wanted Alive: Asian Rhinos in the Wild*. Species Status Report. World Wildlife Fund.

Kenney, J. S., D. Smith, A. M. Starfield, and C. McDougal. 1995. The long-term effects of tiger poaching on population viability. *Conservation Biology* 9 (5): 1127–1133.

Khan, M. 1987. Tigers in Malaysia: Prospects for the future. Pp. 75–84 in *Tigers of the World: The Biology, Biopolitics, Management, and Conservation of an Endangered Species*, edited by R. L. Tilson and U. S. Seal. Noyes.

Khan, M. A. R. 1987. The problem tiger of Bangladesh. Pp. 92–96 in *Tigers of the World: The Biology, Biopolitics, Management, and Conservation of an Endangered Species*, edited by R. L. Tilson and U. S. Seal. Noyes.

Khan, M. K. M., T. J. Foose, and N. van Strien. 2001. Asian Rhino Specialist Group report. *Pachyderm* 31:11–13.

Kipling, R. 1902. *The Just-So Stories*. (Penguin edition, 1998.)

Kitchener, A. C. 1999. Tiger distribution, phenotypic variation and conservation issues. Pp. 19–39 in *Riding the Tiger: Tiger Conservation in Human-Dominated Landscapes*, edited by J. Seidensticker, S. Christie, and P. Jackson. Cambridge University Press.

Bibliography

Kitchener, A. C., and A. J. Dugmore. 2000. Biogeographical change in the tiger, *Panthera tigris*. *Animal Conservation* 3 (2): 113–124.

Klayman, D. L. 1985. Qinghaosu (Artemesinin): An antimalarial drug from China. *Science* 228:1049–1055.

Koreny, F. 1985. *Albrecht Dürer and the Animal and Plant Studies of the Renaissance*. New York Graphic Society.

Koshkarev, E., and V. Vyrypaev. 2001. The snow leopard after the breakup of the Soviet Union. *Cat News* 32:9–11.

Krishnan, M. 2004. On the prowl [Wildlife Poaching]. *India Today* 39 (50):64–66.

Kumar, A., and B. Wright. 1999. Combating tiger poaching and illegal wildlife trade in India. Pp. 243–251 in *Riding the Tiger: Tiger Conservation in Human-Dominated Landscapes*, edited by J. Seidensticker, S. Christie, and P. Jackson. Cambridge University Press.

Kurniawan, D., and R. Nursahid. 2002. Bear Markets: Indonesia. Pp. 93–120 in *The Bear Bile Business: The Global Trade in Bear Products from China to Asia and Beyond*, edited by T. Phillips and P. Wilson. World Society for the Protection of Animals.

Lacey, R., and D. Danziger. 1999. *The Year 1000*. Little, Brown.

Laurie, W. A., E. M. Lang, and C. P. Groves. 1983. *Rhinoceros unicornis* [Indian rhinoceros]. *Mammalian Species* 211:1–6. American Society of Mammalogists.

Leader-Williams, N. 1989. Desert rhinos dehorned. *Nature* 340:599–660.

———. 1992. *The World Trade in Rhino Horn: A Review*. TRAFFIC International.

Leakey, L. S. B. 1969. *The Wild Realm: Animals of East Africa*. National Geographic Society.

Levathes, L. 1994. *When China Ruled the Seas*. Simon and Schuster.

Lewinsohn, R. 1964. *Animals, Men, and Myths*. Premier.

Ley, W. 1948. *The Lungfish, the Dodo, and the Unicorn*. Viking.

———. 1962. *Exotic Zoology*. Viking.

———. 1968. *Dawn of Zoology*. Prentice-Hall.

Li, Y. M., and D. M. Wilcove. 2004. Threats to vertebrate species in China and the United States. *BioScience* 55 (2):137–146

Liao Y., and B. Tan. 1988. A preliminary study on the geographical distribution of snow leopards in China. Pp. 51–63 in *Proceedings of the Fifth Annual Snow Leopard Symposium*, edited by H. Freeman. International Snow Leopard Trust and Wildlife Institute of India.

Linden, E. 1994. Tigers on the brink. *Time* 143 (13): 44–51.

Lindeque, M. 1990. The case for dehorning the black rhinoceros in Namibia. *South African Journal of Science*. 86:226–227.

Lindeque, M., and K. P. Erb. 1996. Research on the effects of temporary horn removal on black rhinos in Namibia. *Pachyderm* 21:27–30.

Linkie, M., D. J. Martyr, J. Holden, A. Yamnuar, A. T. Hartana, J. Sugardjito, and N. Leader-Williams. 2003. Habitat destruction and poaching threaten the Sumatran tiger in Kerinci Seblat National Park, Sumatra. *Oryx* 37 (1): 41–48.

Lloyd, G., and N. Sivin. 2002. *The Way and the Word: Science and Medicine in Early China and Greece*. Yale University Press.

Loutit, B., and S. Montgomery. 1994. The efficacy of rhino dehorning: Too early to tell! *Conservation Biology* 8 (4): 923–924.

Lu, H., 1987. Habitat availability and prospects for tigers in China. Pp. 71–74 in *Tigers of the World: The Biology, Biopolitics, Management, and Conservation of an Endangered Species*, edited by R. L. Tilson and U. S. Seal. Noyes.

Lum, P. 1951. *Fabulous Beasts*. Thames and Hudson.

Lyall, S. 2000. *The Lady and the Unicorn*. Parkstone Press.

Lynam, A. J. 2004. Band of brothers: Soldiers, police, and foresters team up together to save wildlife in Cambodia. *Wildlife Conservation* 107 (4): 22–28.

Lyons, A. S. 1978. Ancient China. Pp. 121–149 in *Medicine: An Illustrated History*, by A. S. Lyons and R. J. Petrucelli. Abrams.

Lyons, A. S., and R. J. Petrucelli. 1978. *Medicine: An Illustrated History*. Abrams.

Ma, Y., and X. Li. 1999. Status and management of the Asiatic black bear in China. Pp. 200–202 in *Bears—Status Survey and Action Plan*, compiled by C. Servheen, S. Herrero, and B. Peyton. IUCN, Gland, Switzerland.

Macdonald, D. 1992. *The Velvet Claw: A Natural History of the Carnivores*. BBC Books.

———, ed. 2001. *The Encyclopedia of Mammals*. Andromeda.

Macilwain, C. 1994. Biologists out of Africa over rhino dispute. *Nature* 368:677.

MacKenzie, D. 1998. First do no harm. *New Scientist* 160 (2157): 53.

Majeski, T. 2002. Bile of black bears may bring a boon for stroke victims. *St. Paul Pioneer Press*, March 15.

Mansfield, A. W., T. G. Smith, and B. Beck. 1975. The narwhal, *Monodon monoceros*, in eastern Canadian waters. *Journal of the Fisheries Research Board of Canada* 32 (7): 1041–1046.

Martin, C. B., and E. B. Martin. 1991. Profligate spending exploits wildlife in Taiwan. *Oryx* 25 (1): 18–20.

Martin, E. B. 1980. Selling rhinos to extinction. *Oryx* 15:322–323.

———. 1981a. Conspicuous consumption of rhinos, part 1. *Animal Kingdom* 84 (1): 11–19.

———. 1981b. Conspicuous consumption of rhinos, part 2. *Animal Kingdom* 84 (2): 20–29.

———. 1983a. Follow-up to stop trade in rhino products in Asia. *Pachyderm* 1:9–12.

———. 1983b. North Yemen bans importation of rhino horn. *Pachyderm* 1:14.

———. 1984. They're killing off the rhino. *National Geographic* 165 (3):404–422.

———. 1985a. Religion, royalty and rhino conservation in Nepal. *Oryx* 19 (1): 11–16.

———. 1985b. Rhinos and daggers: A conservation problem. *Oryx* 19 (4): 198–201.

———. 1987a. Deadly love potions. *Animal Kingdom* 90 (1): 16–21.

———. 1987b. The Yemeni rhino horn trade. *Pachyderm* 8:13–16.

———. 1989a. Report on the trade in rhino products in Eastern Asia and India. *Pachyderm* 11:13–22.

———. 1989b. The rhino product trade in northern and western Borneo. *Pachyderm* 12:38–41.

———. 1990. Medicines from Chinese treasures. *Pachyderm* 13:12–13.

———. 1991. Rhino horn in China: A problem for conservation . . . and the world of art. *Wildlife Conservation* 94 (2): 24–25.

———. 1992a. The poisoning of rhinos and tigers in Nepal. *Oryx* 26:82–86.

———. 1992b. A survey of rhino products for retail sale in Bangkok in early 1992. *Pachyderm* 15:53–55.

———. 1994. Rhino poaching in Namibia from 1980 to 1990 and the illegal trade in the horn. *Pachyderm* 18:39–51.

———. 1995. Tigers in peril in Cambodia. *Oryx* 29 (1): 2–3.

———. 1996. Smuggling routes for West Bengal's rhino horn and recent successes in curbing poaching. *Pachyderm* 21:28–34.

———. 2001. What strategies are effective for Nepal's rhino conservation: A recent case study. *Pachyderm* 31:42–51.

Martin, E. B., L. X. Chen, and C. K. Lin. 1991. The breeding centre for Siberian tigers in China. *International Zoo News* 38 (4): 11–14.

Martin, E. B., and C. B. Martin. 1982. *Run Rhino Run*. Chatto & Windus.

———. 1987. Combatting the illegal trade in rhinoceros products. *Oryx* 21:143–148.

———. 1989. The Taiwanese connection: A new peril for rhinos. *Oryx* 23:76–81.

———. 1991. Profligate spending exploits wildlife in Taiwan. *Oryx* 25:18–20.

Martin, E. B., C. B. Martin, and L. Vigne. 1987. Conservation crisis—the rhinoceros in India. *Oryx* 21:212–218.

Martin, E. B., and T. C. I. Ryan. 1990. How much rhino horn has come into international markets since 1970? *Pachyderm* 13:20–25.

Martin, E. B., and K. H. Smith. 1999. Entrepots for rhino horn in Khartoum and Cairo threaten Garamba's white rhino population. *Pachyderm* 27:76–85.

Martin, E. B., and L. Vigne. 1996. Nepal's rhinos: One of the greatest conservation success stories. *Pachyderm* 21:10–26.

———. 1997. An historical perspective of the Yemeni rhino horn trade. *Pachyderm* 23:29–40.

———. 2003. Trade in rhino horn from eastern Africa to Yemen. *Pachyderm* 34:75–87.

Maruska, E. J. 1987. White tigers: Phantom or freak? Pp. 372–379 in *Tigers of the World: The Biology, Biopolitics, Management, and Conservation of an Endangered Species*, edited by R. L. Tilson and U. S. Seal. Noyes.

Matthiessen, P. 1978. *The Snow Leopard*. Viking.

———. 2000. *Tigers in the Snow*. North Point.

Matyushkin, E. N., O. G. Pikunov, Y. M. Dunishenko, D. G. Miquelle, I. G. Nikolaev, E. N. Smirnov, G. P. Salkina, V. K. Abramov, V. I. Basyllikov, V. G. Yudin, and V. G. Korkishko. 1997. Numbers, distribution and habitat status of the Amur tiger in the Russian Far East. *USAID Russian Far East Environmental Policy and Technology Project* 1–30.

Mazák, V. 1981. Panthera tigris. *Mammalian Species* 152:1–8. American Society of Mammalogists.

McCarthy, Tom. 2000. Snow leopard conservation in Mongolia comes of age. *Cat News* 32:12.

McCarthy, Terry. 2004. Nowhere to roam: Wildlife reserves alone cannot protect big cats. *Time* 164 (8): 44–43.

McDougal, C. 1977. *The Face of the Tiger*. Rivington and Andre Deutsch.

———. 1987. The man-eating tiger in geographical and historical persepctive. Pp. 435–448 in *Tigers of the World: The Biology, Biopolitics, Management, and Conservation of an Endangered Species*, edited by R. L. Tilson and U. S. Seal. Noyes.

———. 1999. Tiger attacks people in Nepal. *Cat News* 30:9.

McDougal, C., A. Barlow, D. Thapa, S. Kumal, and D. B. Tamang. 2004. Tiger and human conflict increase at Chitwan Reserve Buffer Zone, Nepal. *Cat News* 40:3–4.

McGlynn, A. 2003. Man admits to selling endangered pelts. *Quad-City Times*, October 1.

McNeill, W. H. 1997. *Plagues and Peoples*. Anchor Books.

Meacham, C. J. 1997. *How the Tiger Lost Its Stripes*. Harcourt Brace.

Meijaard, E. 1996. The Sumatran rhinoceros in Kalimantan, Indonesia: Its possible distribution and conservation prospects. *Pachyderm* 21:15–23.

Menon, V. 1996. *Under Siege: Poaching and Protection of Greater One-Horned Rhinoceros in India*. TRAFFIC International.

Metcalfe, G. T. C. 1961. Rhinoceros in Malaya and their future. Pp. 183–191 in *Nature Conservation in Western Malaysia*, edited by J. Wyatt-Smith and P. R. Wycherly. Malayan Nature Society.

Miller, J., S. Engelberg, and W. Broad. 2001. *Germs*. Touchstone.

Milliken, T. 1991. South Korea re-visited: The trade in rhino horn and ivory. *Pachyderm* 14:26–28.

Milliken, T., K. Nowell, and J. B. Thomsen. 1993. *The Decline of the Black Rhino in Zimbabwe*. TRAFFIC International.

Mills, J. A. 1992. Milking the bear trade. *International Wildlife* 22 (3): 38–45.

———. 1993. *Market under Cover: The Rhinoceros Horn Trade in South Korea*. TRAFFIC International.

———. 1997. *Rhinoceros Horn and Tiger Bone in China: An Investigation of Trade Since the 1993 Ban*. TRAFFIC International.

———. 1998. Need for further research into tiger bones and musk substitutes agreed. *Traffic Dispatch* April 1998: 5–7.

Mills, J. A., S. Chan, and A. Ishihara. 1995. *The Bear Facts: The East Asian Market for Bear Gall Bladder*. TRAFFIC International.

Mills, J. A., and P. Jackson. 1994. *Killed for a Cure: A Review of Worldwide Trade in Tiger Bone*. TRAFFIC International.

Mills, J. A., and C. Servheen. 1991. *The Asian Trade in Bears and Bear Parts*. TRAFFIC USA/World Wildlife Fund.

Mills, S. 2004. *Tiger*. Firefly.

Milner-Gulland, E. J. 1999. How many to dehorn? A model for decision-making by rhino managers. *Animal Conservation* 2 (2): 137–147.

Milner-Gulland, E. J., J. R. Beddington, and N. Leader-Williams. 1992. Dehorning African rhinos: A model of optimal frequency and profitability. *Proceedings of the Royal Society of London* B 249:83–87.

Milner-Gulland, E. J., O. M. Bukreeva, T. Coulson, A. A. Lushchekina, M. V. Kholodova, A. Bekenov, and Y. A. Grachev. 2003. Reproductive collapse in saiga antelope harems. *Nature* 422:135.

Milner-Gulland, E. J., M. V. Kholodova, A. Bekenov, O. M. Bukreeva, Y. A. Grachev, L. Amgalan, and A. A. Lushchekina. 2001. Dramatic decline in saiga antelope populations. *Oryx* 35 (4): 340–345.

Milner-Gulland, E. J., N. Leader-Williams, and J. R. Beddington. 1994. Is dehorning African rhinos worthwhile? *Pachyderm* 18:52–58.

Mishra, H. M., C. Wemmer, and J. L. D. Smith. 1987. Tigers in Nepal: Management conflicts with human interests. Pp. 449–464 in *Tigers of the World: The Biology, Biopolitics, Management, and Conservation of an Endangered Species*, edited by R. L. Tilson and U. S. Seal. Noyes.

Misra, M. 2000. Pangolin distribution and trade in east and northeast India. *TRAFFIC dispatch* 14:1–4.

Mitchell, E. D., and R. R. Reeves. 1981. Catch history and cumulative catch estimates of initial population size of cetaceans in the eastern Canadian Arctic. *Reports of the International Whaling Commission* 31 (SC/32/O): 645–682.

Montgomery, S. 1995. *Spell of the Tiger: The Man-Eaters of Sundarbans*. Houghton Mifflin.

———. 2001. *The Man-Eating Tigers of Sundarbans*. Houghton Mifflin.

———. 2002. *Search for the Golden Moon Bear: Science and Adventure in Pursuit of a New Species*. Simon and Schuster.

Morgan-Davies, M. 2001. Survey and conservation status of five black rhino (*Diceros bicornis minor*) populations in the Selous Game Reserve, Tanzania, 1997–1999. *Pachyderm* 31:21–35.

Mountfort, G. 1981. *Saving the Tiger*. Viking.

Mulama, M. 2002. Renewed threat to Kenya's conservation efforts. *Pachyderm* 32:85–86.

Nakashima, E. 2003. Thais crack down on wildlife trafficking. *Washington Post*, December 10.

Needham, J. 1981. *Science in Traditional China*. Harvard University Press.

———. 2000. *Science and Civilisation in China*. Vol. 6 of *Biology and Biological Technology*. Part 6, *Medicine*. Cambridge University Press.

Normile, D., and M. Enserink. 2003. Tracking the roots of a killer. *Science* 301:297–299.

Normile, D., and D. Yimin. 2003. Civets back on China's menu. *Science* 301:1031.

Noskova, N. G. 2001. Elasmotherians—evolution, distribution and ecology. 126–128 in *Proceedings of the International Congress: World of Elephants, Rome*. 126–128.

Nowack, R. M. 1991. *Walker's Mammals of the World, Fifth Edition*. Johns Hopkins University Press.

Nowak, R. 2003. SARS could rise again. *New Scientist* 180 (2426/7/8): 15.

Nowell, K. 2000. *Far from a Cure: The Tiger Trade Revisted*. TRAFFIC International.

———. 2002. Revision of the Felidae Red List of Threatened Species. *Cat News* 37:4–7.

Nowell, K.,W.-L. Chyi, and C.-J. Pei. 1992. *The Horns of a Dilemma: The Market for Rhino Horn in Taiwan*. TRAFFIC International.

Nuland, S. D. 1988. *Doctors: The Biography of Medicine*. Knopf.

Nyhus, P. J., Sumianto, and R. Tilson. 1999. Tiger-human interactions in southern Sumatra, Indonesia. Pp. 144–145 in *Riding the Tiger: Tiger Conservation in Human-Dominated Landscapes*, edited by J. Seidendsticker, P. Jackson, and S. Christie. Cambridge University Press.

Nyhus, P. J., R. L. Tilson, and J. L. Tomlinson. 2003. Dangerous animals in captivity: Ex situ tiger conflict and implications for private ownership of exotic animals. *Zoo Biology* 22:573–586.

O'Brien, T. G., M. F. Kinnaird, and H. T. Wibisono. 2003. Crouching tigers, hidden prey: Sumatran tiger and prey populations in a tropical forest landscape. *Animal Conservation* 6 (2): 131–139.

O'Connell-Rodwell, C., and R. Parry-Jones. 2002. *An Assessment of China's Management of Trade in Elephants and Elephant Products*. TRAFFIC East Asia.

O' Hanlon, R. 1984. *Into the Heart of Borneo*. Random House.

O'Neill, H. B. 1987. *Companion to Chinese History*. Facts on File.

Pain, S. 2003. Last of the lions. *New Scientist* 179 (2413): 36–39.

Pala, C. 2003. Operation snow leopard. *Wildlife Conservation* 106 (5): 24–27.

Parker, I., and M. Amin. 1983. *Ivory Crisis*. Chatto & Windus.

Parker, I. S. C., and E. B. Martin. 1979. Trade in African rhino horn. *Oryx* 15:153–58.

Parker, J. T. 2001. The mythic Chinese unicorn *Zhi*. http://www.rom.on.ca/pub/unicorn/index.html (last accessed 2003).

Parry-Jones, R. 2001. Traffic examines musk deer farming in China. *Traffic Dispatches* 16:8.

Pearce, F. 2003. Going the way of the dodo? *New Scientist* 177 (2382): 4–5.

Periera, D., R. Loh, M. B. Bonfiglio, and G. Yung. 2002. Bear markets: Malaysia. Pp. 153–183 in *The Bear Bile Business: The Global Trade in Bear Products from China to Asia and Beyond*, edited by T. Phillips and P. Wilson. World Society for the Protection of Animals.

Perry, R. 1965. *The World of the Tiger*. Atheneum.

Peterson, I. 2002. Cuddly to some, deadly to others; tigers increasingly roam backyards and back streets. *New York Times*, February 1.

Petrucelli, R. J. 1978. The Renaissance. Pp. 369–397 in *Medicine: An Illustrated History*, by A. S. Lyons and R. J. Petrucelli. Abrams.

Phillips, T., and P. Wilson, eds. 2002. *The Bear Bile Business: The Global Trade in Bear Products from China to Asia and Beyond*. World Society for the Protection of Animals.

Pitulko, V. V., P. A. Nikolsky, E. Y. Girya, A. E. Basilyan, V. E. Tumskoy, S. A. Koulakov, S. N. Astakhov, E. Y. Pavlova, and M. A. Anisimov. 2004. The Yana RHS Site: Humans in the Arctic before the last glacial maximum. *Science* 303:52–56.

Pliny. n.d. *Naturalis Historia*. Translated by H. Rackham. 1940. Loeb Classical Library, Harvard University Press.

Plowden, C., and D. Bowles. 1997. The illegal market in tiger parts in northern Sumatra, Indonesia. *Oryx* 31 (1): 59–66.

Polet, G., T. V. Mui, N. X. Dang, B. H. Manh, and M. Baltzer. 1999. The Javan rhinos, *Rhinoceros sondaicus annamiticus*, of Cat Tien National Park, Vietnam: Current status and management. *Pachyderm* 27:34–48.

Bibliography

Potgieter, DeW. 1995. *Contraband: South Africa and the International Trade in Ivory and Rhino Horn*. Queillerie.

Polgreen, L., and J. George. 2003. From a cub to a menace, and now a mystery. *New York Times*, October 6.

Polo, M. n.d. *The Travels*. Translated by R. E. Latham. 1958. Penguin Classics.

Porsild, M. P. 1918. On savssats: A crowding of Arctic animals at holes in the sea ice. *Geographic Review* 6:215–228.

———. 1922. Scattered observations on narwhals. *Journal of Mammalogy* 3:8–13.

Porter, R. 1997. *The Greatest Benefit to Mankind*. Norton.

Prynn, D. 2004. *Amur Tiger*. Russian Nature Press.

Quammen, D. 2003. *Monster of God: The Man-Eating Predator in the Jungles of History and the Mind*. Norton.

Quigley, H. B. 1993. Saving Siberia's tigers. *National Geographic* 184 (1): 38–47.

Rabinowitz, A. 1991. *Chasing the Dragon's Tail: The Struggle to Save Thailand's Wild Cats*. Island Press.

———.1993. Estimating the Indochinese tiger population *Panthera tigris corbetti* in Thailand. *Biological Conservation* 65 (3): 213–217.

———. 1995. Helping a species to go extinct: The Sumatran rhino in Borneo. *Conservation Biology* 9 (3): 482–488.

———. 1999. The status of the Indochinese tiger: Separating fact from fiction. Pp. 148–165 in *Riding the Tiger: Tiger Conservation in Human-Dominated Landscapes*, edited by J. Seidensticker, S. Christie, and P. Jackson. Cambridge University Press.

———. 2001. *Beyond the Last Village*. Island Press.

Rabinowitz, A., G. B. Schaller, and U. Uga. 1995. A survey to assess the status of Sumatran rhinoceros and other large mammal species in Tamanthi Wildlife Sanctuary, Myanmar. *Oryx* 29 (2): 123–128.

Read, B. E. 1931. Chinese materia medica: Animal drugs. *Peking Natural History Bulletin* 5 (4): 37–80 and 6 (1): 1–102.

———. 1934. Chinese materia medica: Dragons and snakes. *Peking Natural History Bulletin* 8 (4):208–362.

Reeves, R. R. 1976. What fate for the narwhal? *North/Nord* 23 (3): 16–21.

Reeves, R. R., and E. Mitchell. 1981. The whale behind the tusk. *Natural History* 90 (8): 50–57.

Reilly, J., G. Hills Spedding, and Apriawan. 1997. Preliminary observations on the Sumatran rhino in Way Kambas National Park, Indonesia. *Oryx* 31 (2): 143–150.

Revkin, A. C. 2003. Hunt imperils polar bears in Bering Sea, report says. *New York Times*, June 17.

———. 2004. White rhino numbers cut by half. *New York Times*, August 7.

Rhode, D. 2003. Rebels in Nepal end cease-fire; government puts army on alert. *New York Times*, August 28.

Ridgeway, R. 2003. Walking the Chang Tang. *National Geographic* 203 (4): 104–123.

Ripley, R. G. 2003. To kill a parasite. *Nature* 424:887–888.

Rodrigues, C. M. P., R. E. Castro, and C. J. Steer. 2004. The role of bile acids in the modulation of apoptosis. *Liver Biology in Health and Disease*, edited by E. E. Bittar, Elsevier Science.

Rodrigues, C. M. P., S. Solä, Z. Nan, R. E. Castro, P. S. Ribeiro, W. C. Low, and C. J. Steer. 2002. Tauoursodeoxycholic acid reduces apoptosis and protects against neurological injury after acute hemorrhagic stroke in rats. *Proceedings of the National Academy of Science* 100 (10): 6087–6092.

Rodrigues, C. M. P., S. Spellman, S. Solä, A. W. Grande, C. Linehan-Stieers, W. C. Low, and C. J. Steer. 2002. Neuroprotection by a bile acid in an acute stroke model in the rat. *Journal of Cerebral Blood Flow & Metabolism* 22:463–471.

Rookmaaker, K. 1997a. Records of the rhinoceros in northern India. *Säugetierkundliche Mitteilungen* 44 (2): 51–78.

———. 1997b. Records of the Sundarbans rhinoceros (*Rhinoceros sondaicus inermis*) in India and Bangladesh. *Pachyderm* 24:37–45.

———. 1998. *The Rhinoceros in Captivity*. SPB Academic Publishing.

———. 2000. The alleged population reduction of the southern white rhinoceros (*Cetrotherium simum simum*) and the successful recovery. *Säugetierkundliche Mitteilungen* 45 (2): 55–70.

———. 2002a. Historical records of the Javan rhinoceros in North-East India. *Rhino Foundation Newsletter* 4:11–12.

———. 2002b. Miscounted population of the southern white rhinoceros in the early 19th century. *Pachyderm* 32:22–27.

———. 2003a. Historical records of the Sumatran rhinoceros in North-East India. *Rhino Foundation Newsletter* 5:11–12.

———. 2003b. Why the name of the white rhinoceros is not appropriate. *Pachyderm* 34:88–92.

———. 2004. Historical distribution of the black rhinoceros (*Diceros bicornis*) in West Africa. *African Zoology* 39 (1): 1–8.

Rookmaaker, L. C., L. Vigne, and E. B. Martin. 1998. The rhinoceros fight in India. *Pachyderm* 25:28–31.

Rose, J. 1972. *Herbs & Things: Jeanne Rose's Herbal*. Perigee.

Roychoudhury, R. 1987. White tigers and their conservation. Pp. 380–388 in *Tigers of the World: The Biology, Biopolitics, Management, and Conservation of an Endangered Species*, edited by R. L. Tilson and U. S. Seal. Noyes.

Saberwal, V. K. 1997. Saving the tiger: More money or less power? *Conservation Biology* 11 (3): 815–817.

Sam, D. D. 1999. Status and management of the Asiatic black bear and sun bear in Vietnam. Pp. 216–218 in *Bears—Status Survey and Action Plan*, compiled by C. Servheen, S. Herrero, and B. Peyton. IUCN, Gland, Switzerland.

Sankhala, K. 1977. *Tiger! The Story of the Indian Tiger*. Simon and Schuster.

Santiapillai, C. 1992. Javan rhinoceros in Vietnam. *Pachyderm* 15:25–27.

Santiapillai, C., and W. S. Ramono. 1987. Tiger numbers and habitat evolution in Indonesia. Pp. 85–91 in *Tigers of the World: The Biology, Biopolitics, Management, and Conservation of an Endangered Species*, edited by R. L. Tilson and U. S. Seal. Noyes.

Sanyal, P. 1987. Managing the man-eaters in the Sundarbans Tiger Reserve of India—a case study. Pp. 427–434 in *Tigers of the World: The Biology, Biopolitics, Management, and Conservation of an Endangered Species*, edited by R. L. Tilson and U. S. Seal. Noyes.

Saper, R. B., S. N. Kales, J. Paquin, M. J. Burns, D. M. Eisenberg, R. B. Davis, and R. S. Phillips. 2004. Heavy metal content of Ayurvedic herbal medicine products. *Journal of the American Medical Association* 292:2868–2873.

Sathyakumar, S. 1999. Status and management of the Asiatic black bear in India. Pp. 202–207 in *Bears—Status Survey and Action Plan*, compiled by C. Servheen, S. Herrero, and B. Peyton. IUCN, Gland, Switzerland.

Scammon, C. M. 1874. *The Marine Mammals of the Northwestern Coast of North America; Together with an Account of the American Whale Fishery*. Carmany and G. P. Putnam's.

Schaller, G. B. 1967. *The Deer and the Tiger*. University of Chicago Press.

———. 1971. Imperiled phantom of Asian peaks. *National Geographic* 140 (5): 702–707.

———. 1979. *Stones of Silence: Journeys in the Himalaya*. Viking.

———. 2003. Drop dead gorgeous: Why poachers are killing the chiru. *National Geographic* 203 (4): 124–125.

Schaller, G. B., N. X. Dang, L. D. Thuy, and V. T. Son. 1990. Javan rhinoceros in Vietnam. *Oryx* 24 (2): 77–80.

Scigliano, E. 2002. *Love, War, and Circuses*. Houghton Mifflin.

Seidensticker, J. 1987a. Bearing witness: Observations on the extinction of *Panthera tigris balica* and *Panthera tigris sondaica*. Pp. 1–8 in *Tigers of the World: The Biology, Biopolitics, Management, and Conservation of an Endangered Species*, edited by R. L. Tilson and U. S. Seal. Noyes.

———. 1987b. Managing tigers in the Sundarbans: Experience and opportunity. Pp. 416–426 in *Tigers of the World: The Biology, Biopolitics, Management, and Conservation of an Endangered Species*, edited by R. L. Tilson and U. S. Seal. Noyes.

———. 1996. *Tigers*. Voyageur.

———. 2002. Tiger tracks. *Smithsonian* 32 (10): 62–69.

Seidensticker, J., S. Christie, and P. Jackson, eds. 1999a. *Riding the Tiger: Tiger Conservation in Human-Dominated Landscapes*. Cambridge University Press.

———. 1999b. Tiger ecology: Understanding and encouraging landscape patterns and conditions where tigers can exist. Pp. 55–60 in *Riding the Tiger: Tiger Conservation in Human-Dominated Landscapes*, edited by J. Seidensticker, S. Christie, and P. Jackson. Cambridge University Press.

Servheen, C. 1999. The trade in bears and bear parts. Pp. 33–38 in *Bears—Status Survey and Action Plan*, compiled by C. Servheen, S. Herrero, and B. Peyton. IUCN, Gland, Switzerland.

Shaw, C. A. 1992. The sabertoothed cats. Pp. 26–27 in *Rancho La Brea: Death Trap and Treasure Trove*, edited by J. M. Harris. Los Angeles Museum of Natural History.

Shepard, O. 1979. *The Lure of the Unicorn*. Harper Colophon.

Shepherd, C. R., and N. Magnus. 2004. *Nowhere to Hide: The Trade in Sumatran Tiger*. TRAFFIC Southeast Asia.

Siderius, C. 2002. Catch those tigers: Years of little or no regulation have made Texas a place where big cats prowl—and sometimes kill. *Dallas Observer*, February 28.

Siebert, C. 2003. Wild thing. *New York Times Magazine*, October 19.

Sillero-Zubiri, C., and D. Gottelli. 1991. Threats to Aberdare rhinos: Predation versus poaching. *Pachyderm* 14:38–39.

Simmons, L. G. 1987. White tigers: The realities. Pp. 389–390 in *Tigers of the World: The Biology, Biopolitics, Management, and Conservation of an Endangered Species*, edited by R. L. Tilson and U. S. Seal. Noyes.

Simon, N. and P. Géroudet. 1970. *Last Survivors*. World.

Singh, N., and H. Lai. 2001. Selective toxicity of dihydroartemisinin and holotransferrin toward human breast cancer cells. *Life Sciences* 70:49–56.

Singh, S. 2003. Enter the tiger. *New Scientist* 180 (2425): 46–49.

Sinha, V. R. 2003. *The Vanishing Tiger*. Salamander.

Siswomartono, D., S. Reddy, W. Ramono, J. Manansang, R. Tilson, N. Franklin, and T. Foose. 1996. The Sumatran rhino in Way Kambas National Park, Sumatra, Indonesia. *Pachyderm* 21:13–14.

Slotow, R., D. Balfour, and O. Howison. 2001. Killing of black and white rhinoceroses by African elephants in Hluhluwe-Umfolozi Park, South Africa. *Pachyderm* 31:14–20.

Smirnov, E. N., and D. G. Miquelle. 1999. Population dynamics of the Amur tiger in Sikhote-Alin Zapovednik, Russia. Pp. 61–70 in *Riding the Tiger: Tiger Conservation in Human-Dominated Landscapes*, edited by J. Seidensticker, S. Christie, and P. Jackson. Cambridge University Press.

Smith, J. L. D., C. McDougal, S. C. Ahearn, A. Joshi, and K. Conforti. 1999. Metapopulation structure of tigers in Nepal. Pp. 176–191 in *Riding the Tiger: Tiger Conservation in Human-Dominated Landscapes*, edited by J. Seidensticker, S. Christie, and P. Jackson. Cambridge University Press.

Smith, J. L. D., S. Tunhikorn, S. Tanhan, S. Simcharoen, and B. Kanchansaka. 1999. Metapopulation structure of tigers in Thailand. Pp. 166–175 in *Riding the Tiger: Tiger Conservation in Human-Dominated Landscapes*, edited by J. Seidensticker, S. Christie, and P. Jackson. Cambridge University Press.

Song, C., and T. Milliken. 1990. The rhino horn trade in South Korea: Still cause for concern. *Pachyderm* 13:6–12.

Spinage, C. A. 1962. *Animals of East Africa*. Houghton Mifflin.

———. 1986. The rhinos of the Central African Republic. *Pachyderm* 6:10–13.

Spitsin, V. V., P. N. Romanov, S. V. Popov, and E. N. Smirnov. 1987. The Siberian tiger (*Panthera tigris altaica*) in the USSR: Status in the wild and in captivity. Pp. 64–70 in *Tigers of the World: The Biology, Biopolitics, Management, and Conservation of an Endangered Species*, edited by R. L. Tilson and U. S. Seal. Noyes.

Sokolov, V. E. 1974. Saiga [*Saiga tartarica*]. *Mammalian Species*. American Society of Mammalogists.

Steer, C. J. 2004. University of Minnesota Faculty Web page. http://www.gcd.med.umn.edu/html/faculty%20pages/steer.html.

Stephan, J. J. 1994. *The Russian Far East: A History*. Stanford University Press.

Stone, R. 2004. A surprsing survival story in the Siberian Arctic. *Science* 303:33.
Stracey, P. D. 1968. *Tigers*. Arthur Barker.
Su, P.-F., J. Wong, and Y. Chiao. 2002. Bear markets: Australia. Pp. 229–238 in *The Bear Bile Business: The Global Trade in Bear Products from China to Asia and Beyond*, edited by T. Phillips and P. Wilson. World Society for the Protection of Animals.
Sunquist, M., and F. Sunquist. 2002. *Wild Cats of the World*. University of Chicago Press.
Sunquist, M., K. U. Karanth, and P. Jackson. 1999. Ecology, behaviour and resilience of the tiger and its conservation needs. Pp. 5–18 in *Riding the Tiger: Tiger Conservation in Human-Dominated Landscapes*, edited by J. Seidensticker, S. Christie, and P. Jackson. Cambridge University Press.
Talbot, R., and R. Whiteman. 1996. *Brother Cadfael's Herb Garden*. Little, Brown.
Taldukar, B. K. 2002. Dedication leads to reduced rhino poaching in Assam in recent years. *Pachyderm* 33:58–63.
Temple, R. 1998. *The Genius of China*. Prion.
Thapar, V. 1992. *The Tiger's Destiny*. Kyle Cathie.
———. 1999a. The tragedy of the Indian tiger: Starting from scratch. Pp. 296–306 in *Riding the Tiger: Tiger Conservation in Human-Dominated Landscapes*, edited by J. Seidensticker, S. Christie, and P. Jackson. Cambridge University Press.
———. 1999b. *The Secret Life of Tigers*. Oxford University Press.
———. 2000. *Wild Tigers of Ranthambhore*. Oxford University Press.
———. 2003. The status of the tiger in India. *Cat News* 39:21–22.
Theile, S. 2003. *Fading Footprints: The Killing and Trade of Snow Leopards*. TRAFFIC International.
Tilson, R. L., and P. J. Nyhus. 1998. Keeping problem tigers from becoming problem species. *Conservation Biology* 12 (2): 261–262.
Tilson, R. L., and U. S. Seal, eds. 1987. *Tigers of the World: The Biology, Biopolitics, Management, and Conservation of an Endangered Species*. Noyes.
Tilson, R. L., K. Traylor-Holzer, and M. J. Qiu. 1997. The decline and impending extinction of the South China tiger. *Oryx* 31 (4): 243–252.
Togawa, K., and M. Sakamoto,. 2002. Bear Markets: Japan. Written with the assistance of C. Iijima. Pp. 121–152 in *The Bear Bile Business: The Global Trade in Bear Products from China to Asia and Beyond*, edited by T. Phillips and P. Wilson. World Society for the Protection of Animals.
Toon, A., and S. Toon. 2002. *Rhinos: Natural History and Conservation*. World Life Library/Voyageur Press.
Topsell, E. 1658. *The Historie of Foure-Footed Beastes*. E. Coates, London.
TRAFFIC. 2004. Armoured but endangered. *Asian Geographic*. 4:64–71.
't Sas-Rolfes, M. 1997. Elephants, rhinos and the economics of the illegal trade. *Pachyderm* 24:23–29.
Tucker, J. B. 2001. *Scourge: The Once and Future Threat of Smallpox*. Atlantic Monthly Press.
Turner, A. 1997. *The Big Cats and Their Fossil Relatives*. Columbia University Press.
Veith, I., trans. 2002. *The Yellow Emperor's Classic of Internal Medicine*. University of California Press.

Vennerstrom, J. L., S. Arbe-Barnes, R. Brun, S. A. Charman, F. C. K. Chiu, J. Chollet, Y. Dong, A. Dorn, D. Hunziker, H. Matile, K. McIntosh, M. Padmanilayam, J. S. Tomas, C. Scheurer, B. Scorneaux, Y. Tang, H. Urwyler, S. Wittlin, and W. N. Charman. 2004. Identificaton of an antimalarial synthetic trioxolane drug development candidate. *Nature* 430:900–904.

Verney, P. 1979. *Homo Tyrannicus: A History of Man's War Against Animals*. Mills & Boon.

Vervoordt, A. 2002. *Catalog 2*. Antwerp.

Vigne, L., and E. B. Martin. 1989. Taiwan: The greatest threat to the survival of Africa's rhinos. *Pachyderm* 11:23–25.

———. 1991. African and Asian rhino products for sale in Bangkok. *Pachyderm* 14:41–43.

———. 1998. Dedicated field staff continue to combat rhino poaching in Assam. *Pachyderm* 26:25–39.

———. 2000. Price for rhino horn increases in Yemen. *Pachyderm* 28:91–100.

Vincent, A. C. J. 1994. The improbable seahorse. *National Geographic* 186 (4): 126–140.

———. 1996. *The International Trade in Seahorses*. TRAFFIC International.

Wallace, A. R. 1869. *The Malay Archipelago: The Land of the Orangutan and the Bird of Paradise*. Macmillan. (Dover edition, 1962.)

Wang, Y. 1999. Status and management of the Formosan black bear in Taiwan. Pp. 213–215 in *Bears—Status Survey and Action Plan*, compiled by C. Servheen, S. Herrero, and B. Peyton. IUCN, Gland, Switzerland.

Ward, G. C. 1997. Making room for wild tigers. *National Geographic* 192 (6): 2–35.

Ward, G. C., and D. R. Ward. 1993. *Tiger-Wallahs*. HarperCollins.

Ward, P. D., and S. Kynaston. 1995. *Bears of the World*. Blandford.

Wemmer, C., J. L. D. Smith, and H. R. Mishra. 1987. Tigers in the wild: The biopolitical challenges. Pp. 396–405 in *Tigers of the World: The Biology, Biopolitics, Management, and Conservation of an Endangered Species*, edited by R. L. Tilson and U. S. Seal. Noyes.

Wendt, H. 1959. *Out of Noah's Ark*. Houghton Mifflin.

Western, D. 1989. The undetected trade in rhino horn. *Pachyderm* 11:26–29.

Western, D., and L. Vigne. 1984. The status of rhinos in Africa. *Pachyderm* 4:5–6.

Whaley, F. 2004. Project Seahorse: Tackling the traditional medicine trade in China. *Wildlife Conservation* 107 (4): 28–33.

Whitaker, B. 2003. Many dead tigers are found at big cat "Retirement Home." *New York Times*, April 24.

White, T. H. 1954. *The Book of Beasts*. Jonathan Cape.

Wikramanayake, E. D., E. Dinerstein, J. G. Robinson, K. U. Karanth, A. Rabinowitz, D. Olson, T. Mathew, P. Hedao, M. Connor, G. Hemley, and D. Bolze. 1997. Where can tigers live in the future? A framework for identifying high-priority areas for the conservation of tigers in the wild. Pp. 255–272 in *Riding the Tiger: Tiger Conservation in Human-Dominated Landscapes*, edited by J. Seidensticker, S. Christie, and P. Jackson. Cambridge University Press.

Bibliography

Wilson, E. O. 2001. Biodiversity: Wildlife in trouble. Pp. 18–20 in *The Biodiversity Crisis: Losing What Counts*, edited by M. J. Novacek. American Museum of Natural History.

———. 2002. *The Future of Life*. Knopf.

Winchester, S. 2003. *Krakatoa: The Day the World Exploded, August 27, 1888*. HarperCollins.

Wines, M. 2003. Zimbabwe's woes are bringing grief to its wildlife, too. *New York Times*, October 25.

Wiseman, N., and A. Ellis, trans. 1996. *Fundamentals of Traditional Chinese Medicine*, rev. ed. Paradigm.

Wright, B. 2004. Spate of tiger and leopard skin seizures. *Cat News* 41:19.

Wu, C., comp. 2002. *Basic Theory of Traditional Chinese Medicine*. Shanghai University of Traditional Chinese Medicine.

Yardley, J. 2004a. Rats hunted in SARS episode in China: New case is confirmed. *New York Times*, January 8.

———. 2004b. The SARS scare in China: Slaughter of the animals. *New York Times*, January 7.

———. 2004c. W.H.O. urges China to use caution while killing civet cats. *New York Times*, January 6.

Yi, H., C. Chen, P. Tohill, S. Smith, and K. Schenkel. 2002. Bear markets: North America (USA and Canada). Pp. 77–92 in *The Bear Bile Business: The Global Trade in Bear Products from China to Asia and Beyond*, edited by T. Phillips and P. Wilson. World Society for the Protection of Animals.

Zhang, E., ed. 1989. *Rare Chinese Materia Medica*. Shanghai University of Traditional Chinese Medicine.

———. 1990. *The Chinese Materia Medica*. Shanghai University of Traditional Chinese Medicine.

Illustration Credits

Frontispiece. Photograph by Richard Ellis

7	Photograph by Richard Ellis
14	Drawing by Richard Ellis
18	Drawing by Richard Ellis
22	Drawing by Richard Ellis
24	Drawing by Richard Ellis
68	Photograph by Richard Ellis
72	From Conrad Gesner's 1551 *Historia Animalium*
75	Photo courtesy of French Ministry of Culture and Communications, Regional Direction for Cultural Affairs Rhone-Alpes Region Regional Department of Archaeology.)
77	Drawing by Richard Ellis
82	Photograph: Metropolitan Museum of Art
84	Photograph by Fred Breummer
87	Photograph: Christie's Images
89	From Conrad Gesner's 1551 *Historia Animalium*
91	Drawing by Richard Ellis
92	Photograph: American Museum of Natural History

Illustrations

Illustrations

Index

Index

Index

Index

Index

Index

Index

Index

Index

Index